MULTILEVEL CONVERTERS

FOR INDUSTRIAL APPLICATIONS

INDUSTRIAL ELECTRONICS SERIES

Series Editors:
Bogdan M. Wilamowski & J. David Irwin

PUBLISHED TITLES

Multilevel Converters for Industrial Applications, *Sergio Alberto González, Santiago Andrés Verne, and María Inés Valla*

Granular Computing: Analysis and Design of Intelligent Systems, *Witold Pedrycz*

Industrial Wireless Sensor Networks: Applications, Protocols, Standards, and Products, *V. Çağrı Güngör and Gerhard P. Hancke*

Power Electronics and Control Techniques for Maximum Energy Harvesting in Photovoltaic Systems, *Nicola Femia, Giovanni Petrone, Giovanni Spagnuolo, and Massimo Vitelli*

Extreme Environment Electronics, *John D. Cressler and H. Alan Mantooth*

Renewable Energy Systems: Advanced Conversion Technologies and Applications, *Fang Lin Luo and Hong Ye*

Multiobjective Optimization Methodology: A Jumping Gene Approach, *K.S. Tang, T.M. Chan, R.J. Yin, and K.F. Man*

The Industrial Information Technology Handbook, *Richard Zurawski*

The Power Electronics Handbook, *Timothy L. Skvarenina*

Supervised and Unsupervised Pattern Recognition: Feature Extraction and Computational Intelligence, *Evangelia Micheli-Tzanakou*

Switched Reluctance Motor Drives: Modeling, Simulation, Analysis, Design, and Applications, *R. Krishnan*

FORTHCOMING TITLES

Smart Grid Technologies: Applications, Architectures, Protocols, and Standards, *Vehbi Cagri Gungor, Carlo Cecati, Gerhard P. Hancke, Concettina Buccella, and Pierluigi Siano*

Data Mining: Theory and Practice, *Milos Manic*

Electric Multiphase Motor Drives: Modeling and Control, *Emil Levi, Martin Jones, and Drazen Dujic*

Sensorless Control Systems for AC Machines: A Multiscalar Model-Based Approach, *Zbigniew Krzeminski*

Next-Generation Optical Networks: QoS for Industry, *Janusz Korniak and Pawel Rozycki*

Signal Integrity in Digital Systems: Principles and Practice, *Jianjian Song and Edward Wheeler*

FPGAs: Fundamentals, Advanced Features, and Applications in Industrial Electronics, *Juan Jose Rodriguez Andina and Eduardo de la Torre*

Dynamics of Electrical Machines: Practical Examples in Energy and Transportation Systems, *M. Kemal Saioglu, Bulent Bilir, Metin Gokasan, and Seta Bogosyan*

MULTILEVEL CONVERTERS
FOR INDUSTRIAL APPLICATIONS

Sergio Alberto González
Santiago Andrés Verne
María Inés Valla

CRC Press
Taylor & Francis Group
Boca Raton   London   New York

CRC Press is an imprint of the
Taylor & Francis Group, an **informa** business

CRC Press
Taylor & Francis Group
6000 Broken Sound Parkway NW, Suite 300
Boca Raton, FL 33487-2742

First issued in paperback 2017

ISBN 13: 978-1-138-07649-5 (pbk)
ISBN 13: 978-1-4398-9559-7 (hbk)

Library of Congress Cataloging-in-Publication Data

González, Sergio Alberto.
    Multilevel converters for industrial applications / Sergio Alberto González, Santiago Andrés Verne, María Inés Valla.
        pages cm. -- (Industrial electronics series)
    Includes bibliographical references and index.
    ISBN 978-1-4398-9559-7 (hardback)
    1. Electric current converters. I. Verne, Santiago Andrés. II. Valla, María Inés. III. Title.

TK7872.C8G65 2013
621.3815'322--dc23   2013018222

Visit the Taylor & Francis Web site at
http://www.taylorandfrancis.com

and the CRC Press Web site at
http://www.crcpress.com

Contents

Preface

In this book, a thorough and comprehensive analysis of multilevel converters with a common DC voltage source is presented. Their control and modulation strategies are described as well as their characteristics and design trade-offs.

The fact that power electronic converters are at the core of modern conversion systems, together with the need for improving efficiency and operation flexibility, lead to constant technical challenges around power converter topologies and control methods. Nowadays, modern semiconductor devices have reached high current and voltage levels. In addition, their power handling limits can be extended if they are used in multilevel converter configurations. However, a detailed understanding of characteristics and the operation of these topologies is mandatory in order to carry out high-performance and reliable control designs. This book is intended to provide deep insight into high-power multilevel converters. It serves as a reference for academic researchers, graduate students, and practicing engineers who are interested in medium-voltage power conversion. These types of converters are increasingly applied in several industries as well as in renewable energy and distributed generation, and other subjects related to the state of the art of power processing systems.

The book is organized into seven chapters.

Chapter 1 provides an overview of medium-voltage power converters and their applications. The next four chapters are dedicated to the analysis of the different multilevel converters, and the last two chapters are dedicated to the analysis of two important case studies, reactive and harmonic compensation and medium-voltage motor drive.

Chapter 2 describes the generalized multilevel converter topology from which the classic converters with a common DC bus are derived, that is, the diode-clamped multilevel converter and the flying capacitor multilevel converter. The generalized topology together with a new graphic representation of the different states of the converter and their voltage levels is employed to analyze the common characteristics of the symmetric topologies.

Chapter 3 analyzes the operation of the diode-clamped multilevel converter, and a multilevel space vector modulation is described. Some internal operating constraints of the converter are also explained and addressed in order to achieve reliable converter operation. Finally, the balancing boundary of the passive front-end converter is characterized.

Chapter 4 describes the operation of the flying capacitor multilevel converter. The dynamic behavior of the inner capacitors is analyzed with a time domain model to understand their self-balance through the phase-shifted carrier pulse width modulation and the use of a tuned balancing network.

Chapter 5 presents a new asymmetric topology with hybrid modulation and a common DC source called a cascade asymmetric multilevel converter (CAMC) with five voltage levels. This topology reaches a higher ratio between the obtainable voltage levels and the amount of converter states. The advantages of CAMC, when compared to more classical topologies, are shown.

Chapter 6 presents a case study that analyzes the behavior of the CAMC as a distribution static compensator (DSTATCOM) and shunt active power filter. This is one of the main applications of multilevel converters to improve power quality in medium-voltage distribution systems as custom power devices.

Chapter 7 presents a case study that analyzes the behavior of the diode-clamped topology configured as a back-to-back converter. In this case, a predictive control approach is described for both the grid-side and load-side converters. The converters and control behavior are analyzed for several working conditions.

About the Authors

Sergio Alberto González, Ph.D., received electronics engineer, master in engineering, and doctor in engineering degrees from the National University of La Plata (UNLP), Argentina, in 1992, 2000, and 2010, respectively. He is full professor of power electronics at UNLP. Also, he is an associate professor of power electronics and motors control at the National University of Quilmes (UNQ), Argentina (since 2000).

Dr. González joined the Industrial Electronics, Control and Instrumentation Laboratory (LEICI) at UNLP in 1992. His research interests have been in the field of power converters, in particular DC-DC converters, resonant converters, and multilevel converters and their application in flexible AC transmission and power quality control.

Santiago Andrés Verne, Ph.D., received electronics engineer and doctor in engineering degrees from the National University of La Plata (UNLP) in 2003 and 2012, respectively. He has been with the Industrial Electronics, Control and Instrumentation Laboratory (LEICI) at UNLP, since 2003, studying multilevel converters and drives. Dr. Verne is currently a head teaching assistant within the Electrical Engineering Department at UNLP.

María Inés Valla, Ph.D., received electronics engineer and doctor in engineering degrees from the National University of La Plata (UNLP) in 1980 and 1994, respectively. She is a full professor in the EE Department at UNLP. She is also a member of the National Research Council of Argentina (CONICET). Dr. Valla joined the Industrial Electronics, Control and Instrumentation Laboratory (LEICI) at UNLP in 1980, and since 1998 she has been the head of the power electronics group within LEICI. Her research interests are in the field of power electronics and AC drives.

Dr. Valla joined IEEE in 1979 as a student member. She has served in different positions, including member of the Fellows Committee (2012–2013), member of the Ethics and Membership Conduct Committee (2006–2008), coeditor in chief of *IEEE Transactions on Industrial Electronics*

(2013–Present), associate editor of *IEEE Transactions on Industrial Electronics* (2007–2012), vice president for membership activities of the Industrial Electronics Society (IES) (2010–2012), and life member of the Administrative Committee of IES (since 2010).

Dr. Valla has been an IEEE fellow since 2010 and a member of the Buenos Aires Academy of Engineering in Argentina since 2007.

chapter 1

Introduction

1.1 Introduction

Power electronics is a discipline increasingly involved in all processing stages of electric power, like generation, conversion, transmission distribution, and conditioning. But the power converters are restricted in their operational capacities by the switching devices, whose limitations are imposed by the physical characteristics of the semiconductor materials. In this sense, large amounts of research are taking place around the development of new semiconductor switching devices with larger voltage withstanding capabilities. However, the aim of increasing the working voltage of the converters with the existing power switches also finds its own way with the introduction of multilevel converters. These converters also have interesting additional features when compared to the classical two-level topologies, such as reducing the harmonic content of the synthesized voltage waveform. However, as the multilevel topologies increase the working voltage of the converters, they greatly increase the number of electronic devices or complementary energy storage elements (capacitors) in the circuit. This implies that the dynamics inside the converters have to be taken into account with their corresponding constraints, whereby the control strategies should consider them as part of the whole control target. Moreover, as the number of switches increases, so do the possible switching combinations, and more elaborated switching algorithms are necessary to meet the multiple constraints of the control problem (number of switches/number of redundant states, power density, unbalanced losses, reliability, etc.). On the other hand, in addition to the complexity of the converter's control, today's modern digital processing platforms allow the practical implementation of sophisticated control strategies at high speed.

All this represents a rich field of research around the study of existing multilevel topologies, the synthesis of new circuit configurations and control algorithms, as well as the dynamic interaction between the converters and the external systems in a wide variety of applications, such as electrical power conversion and conditioning.

1.2 Medium-Voltage Power Converters

The increasing energy demand encourages researchers to devise more efficient power processing systems. In this sense, electronic converters have gained significant prominence due to their versatility and fast dynamic response. Also, the development of advanced control techniques, and specially the technological evolution of power electronic switches, has made possible the operation of power converters in the medium-voltage range.

Nowadays, high-power converters can be found in numerous applications, such as AC drives for process control in the petrochemical, mining, and steel industries and also in traction technology for transportation [1–12]. The power processing in electrical networks at the distribution level for power quality improvement and also for the optimization of existing electrical infrastructure is a profitable application field [10,13–17]. Closely related to this field is the necessity of a grid-friendly interface of the renewable energy sources. In this discipline, the power converters are a key component because the control problems involve multiple and severe constraints from the grid and from the energy resource that can only be addressed through the highly versatile electronic converters.

The advancement of the gate turn-off thyristor (GTO) [18,25,26] and its evolution, the integrated gate-commutated thyristor (IGCT) [19–24], have given great impulse to the voltage source converters (VSCs) at medium voltage. This type of high-power semiconductor is more appropriate to be used in classical two-level VSCs. Meanwhile, the technology of the insulated gate bipolar transistor (IGBT) has given a strong push into the study of converters with multilevel topologies [8,23].

Technological evolution of power switches has allowed the development of motor drives up to 100 MW and static power compensators of about 10 MVA. Multilevel topologies have joined this evolution, ranging from 2.2 to 6.6 kV without the use of coupling transformers [10]. Moreover, they present some additional advantages over classical two-level topology, like the reduction of voltage harmonics and its consequent improvement of the processed power quality and also the increment of power density of the converters.

Nowadays, there is great interest around the study of voltage source multilevel converters (VSMCs), especially regarding new converter topologies and control methods. In this sense, different topologies were developed as industrial products with sinusoidal modulation techniques [26,27]. These techniques, which are well known for two-level converters, have been extended to multilevel converters [6,15,24,40,41], as well as a wide variety of advanced nonlinear control strategies, such as predictive control, among others.

1.3 Multilevel Converters

The AC output voltage of the two-level VSC is limited by the maximum voltage blocking capability of its switching devices. The use of VSC in medium-voltage applications without coupling transformers is severely constrained. So, a widespread technique to overcome these limitations consists of the series connection of the power devices, which allows distributing the blocking voltage [21]. Figure 1.1a shows one leg of a classical two-level VSC, whose switches are implemented with the series connection of two power devices. Considering that the series-connected devices have identical static and dynamic characteristics, the blocking voltage will be equally shared and each device should withstand $V_{DC}/2$. This concept can be generalized to an arbitrary number of switching devices and allows an unlimited extension of the output voltage. However, this is not completely true due to natural spread of the electrical characteristics of the switching devices and their driving circuits. This introduces static and dynamic unbalances on the blocking voltages on each device during commutation. Therefore, the use of external snubber networks becomes essential to equalize the blocking voltages statically and dynamically. Integrated switching blocks and driving circuits with active overvoltage

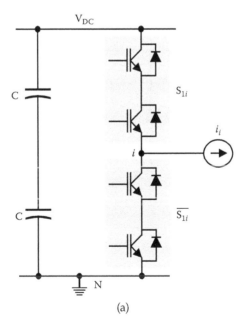

(a)

Figure 1.1 (a) Two-level converter with series-connected switches. (b) Neutral point clamp converter (three-level). (c) Flying capacitor converter (three-level).

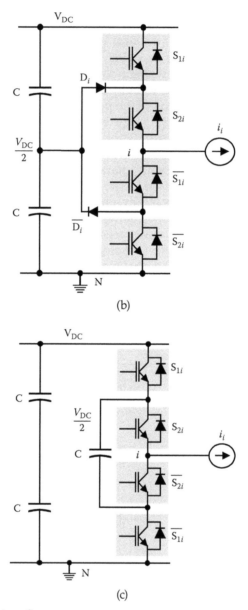

(b)

(c)

Figure 1.1 (Continued)

clamping and protections have also been developed in order to improve switching features [22]. However, they still have some drawbacks, like low utilization factor of the power semiconductors, increase of the switching losses, and need to duplicate or triplicate the redundancy of the blocks to handle the fault situation.

The necessity to equalize the blocking voltage on the devices connected in series gave origin to the VSMC [4]. The first development in multilevel topologies has been the neutral point clamped (NPC) converter [28]. Figure 1.1b shows one leg of the NPC converter with four active switches that are connected to the midpoint of the DC bus through the clamping diodes (D_i and $\bar{D}_i$).

Each power semiconductor is an individual switch, and the upper switches ($S_{(1,2)i}$) are activated in a complementary way with respect to those at the bottom of the leg ($\bar{S}_{(1,2)i}$). The effect of voltage limiting over the active switches is due to the clamping diodes D_i and $\bar{D}_i$, which fix the blocking voltage to $V_{DC}/2$. Another multilevel topology that allows voltage clamping on the power switches is the flying capacitor multilevel converter (FCMC) introduced by Meynard and Foch [29]. Figure 1.1c shows one leg of this converter where the flying capacitor is connected between the upper switches pair and bottom switches pair. The flying capacitor is supposed to have a constant voltage equal to the half of the DC bus voltage. Therefore, $V_{DC}/2$ is the blocking voltage of all switches of the leg. However, it is necessary to implement an adequate switching sequence in order to keep the voltage of the flying capacitor [30].

A comparison between the topologies of Figure 1.1 shows that they have the same number of power semiconductors (without considering the clamping diodes of the NPC converter). However, they do not have the same number of switches. While the two-level VSC leg has two switches, both the NPC converter and FCMC topologies have four switches. The additional switches and their interconnection nodes are the key to increase the number of levels on the output voltage waveform. The classical two-level topology allows us to define only two possible voltages between nodes i and N (0 and V_{DC}), and both the NPC converter and FCMC topologies are able to generate three voltage levels between nodes i and N: 0, $V_{DC}/2$, and V_{DC}. The appearance of an intermediate voltage level ($V_{DC}/2$) allows reducing the harmonic contents on the output voltage. Hence, the multilevel topologies not only provide a clamping mechanism for the power switches, but also allow improving the harmonic quality of the synthesized output voltage.

Another voltage source multilevel converter that has interesting applications is the cascaded cell multilevel converter (CCMC) [7,24,31–33]. The simplest topology of the CCMC is shown in Figure 1.2 (five levels). It comprises two series-connected H-bridges. Each bridge imposes three voltage levels on the AC terminals (0, $\pm V_{DC}/2$), whereupon the switching combinations of both H-bridges yield to five output voltages: 0, $\pm V_{DC}/2$, and $\pm V_{DC}$.

1.3.1 Symmetric Topologies

All the VSMCs presented in the previous section have symmetric topologies. This feature allows expanding the topology to an arbitrary number

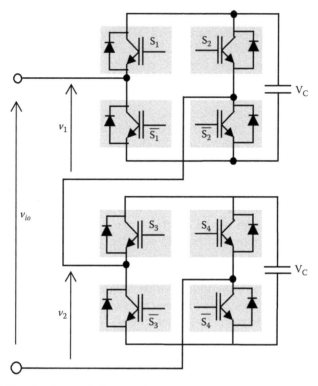

Figure 1.2 Five-level cascaded converter.

of levels by stacking basic cells, as demonstrated by Choi et al. [5], Lai and Pen [31], Peng [32], and Rodriguez et al. [33]. An *n*-level symmetric structure requires $2(n-1)$ active devices in each converter leg. The *n*-level diode-clamped multilevel converter (DCMC), which is the generalized version of the NPC converter, requires $(n-1)$ series-connected capacitors on the DC bus and $(n-1).(n-2)$ equally rated diodes [34]. On the other hand, the *n*-level FCMC does not need any clamping diode. However, it requires $(n-1).(n-2)/2$ flying capacitors in each leg of the converter. The number of levels (*n*) of the symmetric topologies can be easily increased adding stages. So, the quality of the voltage waveform and the DC voltage of the converter are increased. Nevertheless, the required number of devices and assembly complexity becomes impractical when the number of voltage levels grows [1]. This is clearly seen in commercial implementations, which generally do not exceed $n = 4$ [3,36–38].

In addition, each topology presents particular disadvantages. For example, the DCMC, with more than three levels, suffers a natural voltage unbalancing on the DC bus capacitors, which requires extra hardware or intelligent control of the switching sequences in order to counteract this effect [37,39]. Another disadvantage is the higher conduction losses of the inner devices with respect to those located at the extremes of the converter leg because it leads to component oversizing and cooling asymmetry [40]. Meanwhile, FCMCs are commercially found with a maximum of four levels. This is almost exclusively due to the large number of capacitors and the associated volume and cost that greatly influence the total cost of the converter [41].

Regarding the CCMC, it can be mentioned that the phase voltage, with five or more odd levels, is straightforward synthesized. However, its main disadvantage lies in the need of isolated DC voltage sources for each bridge. Moreover, it is impossible to build back-to-back connections with this topology.

1.3.2 Asymmetric Topologies

A variant of the CCMC topology is the hybrid multilevel converter (HMC), which has the same structure as the CCMC but different DC voltage sources on each H-bridge [31,43–46]. In this way, the number of voltage levels are increased, keeping the same number of switching devices. This is a very interesting approach since it offers an improvement in the output voltage quality without raising the system complexity. In particular, the adoption of an adequate progression of the DC voltages jointly with a hybrid modulation strategy allows a significant optimization of the ratio between the number of levels and the number of switching devices. This has led to a design tendency that searches for multilevel topologies that are able to synthesize multiple voltage levels with a low number of switching devices. The literature presents several combinations of H-bridge cells with NPC converter structures, among others [47,48], but an important disadvantage persists: due to the lack of a common DC bus, these topologies cannot be connected in back-to-back configuration. In addition, they still require isolated DC sources for each cell of the converter.

The need for multilevel topologies with a higher number of levels, a reduced number of switching devices, and a common DC bus has given rise to new hybrid topologies [9,48]. However, few studies suggest the idea to reduce the converter redundant states such that each voltage level is generated from a unique switching combination [49]. The analysis and development of new asymmetric converter configurations, like the one presented by Barbosa et al. [48], led to the development of the cascade asymmetric multilevel converter (CAMC) [50,51]. With both the cascaded asymmetric structure and the hybrid modulation scheme, it is possible to synthesize a five-level voltage waveform. It presents a significant improvement in terms of the relationship between the number of voltage levels and the number of switching states.

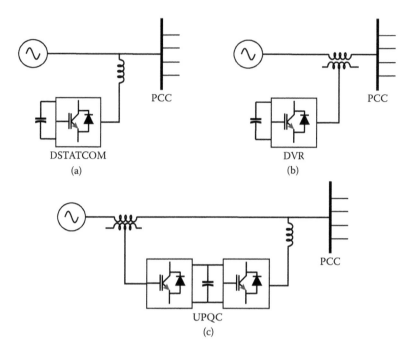

Figure 1.3 Connection topologies for custom power devices: (a) DSTATCOM, (b) DVR, and (c) UPQC.

1.4   Applications

1.4.1   Power Quality Improvement

Electrical grids are subject to several types of disturbances that degrade the power quality (PQ) of the power systems [13–15]. Depending on their magnitude, critical consumers may be significantly affected. The degrading of the PQ can be characterized as voltage transients, sags, swells, interruptions, voltage distortions, flickers, and voltage unbalances, among others.

The custom power devices (CPDs) have been designed to improve the PQ of the electrical system [13]. Some of these devices can be implemented with VSMC. There exist different connection modes of the compensation systems to the grid, which can address more effectively each particular PQ disturbance. There are three main connection topologies. Figure 1.3 shows the three types of compensators acting at the point of common coupling (PCC):

- *Shunt connection* [42,52]: When utilized in distribution grids, it is commonly called distribution static compensator (DSTATCOM). Its main capabilities are reactive power compensation, voltage regulation, flicker mitigation, harmonic current filtering, and load balancing. It may include an energy storage device for a short temporary fault of the utility grid.

- *Series connection* [53–55]: It is called dynamic voltage restorer (DVR). It is generally used for voltage stabilization of critical consumers. It frequently has an energy storage element for a short voltage loss of the main power source.
- *Combined shunt and series connection* [56]: It is commonly called universal power quality conditioner (UPQC). It is the most flexible choice among all custom power devices (and the most expensive) since it puts together all the advantages of both previously mentioned CPDs. Its structure comprises two VSMCs in back-to-back connection; one of them is connected in series with the load and the other is connected in parallel.

1.4.2 Renewable Energy Interconnection

Another field of increasing research and involvement of multilevel converters corresponds to renewable energy systems, primarily motivated to address interconnection issues between alternative power sources like solar, wind, or hydrogen and the electrical networks. A versatile electronic power interface between the source and the grid allows us to optimize the energy extraction with a simultaneous control of several system variables, offering high-power quality standards and even ancillary services. This imposes challenges on the control of grid parameters, and in this sense the VSMC appears as a key component of the control systems, especially in the MV range [57,58].

The race for lower costs of generated energy is a great incentive for wind turbine manufacturers to develop more powerful and constantly growing machines. An exponential growth of the generation capacity can be observed from the beginning of the new era of wind power generation in the 1970s. The size and power of commercial wind turbines has greatly increased in the last years. Nowadays, the industry standard ranges from 1 to 2 MW, and modern designs reach 6 to 7.5 MW; such is the case of the newest wind turbine by Enercon (Model E-126).

Modern wind turbines operate at variable speed due to better energy production with respect to their constant speed counterparts. Several combinations exist between different types of generators and power converters in both commercial products and research areas [59,60]. One of the most popular designs is based on the doubly fed induction generator (DFIG) with reduced power rotor converter. This topology has a controllability margin of the rotor speed that ranges ±30%, which is a widely accepted value for the optimization of the annual energy production. The percentage of speed regulation with respect to the rated speed is the same as the ratio of the rated power of the rotor converter related to the output power of the turbine. Although this reduction is, at first, an interesting feature in normal operation, it limits the operational capacity of the system when ancillary services are demanded by the grid operators.

A promising topology that has won considerable interest in the last few years is based on synchronous generators injecting power to the grid through a full-scale power converter. In this case, the converter extracts the mechanical energy from the generator at variable voltage and variable frequency and injects it to the grid at constant voltage and frequency. As the power converter is connected to the grid with 100% operational capacity, ancillary services can be provided in the same scale of the system rated power, bringing full support to the accomplishment of modern grid codes [61,62]. A particular type of machine that is also promising in this category is the permanent magnet synchronous generator. In the last few years, the decreasing cost of modern and NdFeB magnets made it viable to build large generators with this technology [60,63,64]. It offers important maintenance benefits because it dispenses the slip rings and also increases the efficiency due to the lack of rotor excitation losses. A major feature also appears because it is possible to build machines with a large number of poles, reducing the rotor speed, and consequently the number of stages of turbine gearbox, thus increasing the efficiency and robustness of the whole system [35].

In the megawatts range, the power generation and its injection to the grid become convenient at medium-voltage levels, in comparison with the parallel connection of low-voltage converters and coupling transformers. Here, the use of multilevel converters also arises as a promising technology, considering the increasing penetration of wind energy on the electric power network.

1.4.3 Variable Speed Drives

High-power AC drives are a major topic in industrial environment, and also the VSMCs are on the scene. Driving pumps for pipeline transport in the oil industry, water and wastewater treatment, high-power blowers in the cement industry, mining machinery, and traction equipment for rail and naval transportation require high levels of power conversion. On the other hand, the need to improve versatility and performance of power conversion leads to the development of power electronics converters with intelligent control strategies [6,65]. The design of an electronic power converter for motor drive applications should consider several issues regarding control requirements and power quality on both sides of the converter (line and load sides) and, simultaneously, implementation and reliability features. As the conversion power reaches several megawatts, the operation at medium voltage up to 10 kV becomes an advantageous choice. However, several issues that are not significant for low-power drives become of great importance when operating at medium-voltage levels, and additional considerations come into play [6].

In high-voltage and power levels, several features and technical issues of the electric drives have to be considered. For example, bidirectional power transfer capability or electromagnetic braking is commonly required when a high amount of energy is stored in the rotational masses. Other issues are related with the quality of the power drained from the grid by the drive converter, like harmonic contamination or reactive power consumption. The interaction between the converter and the machine also presents important issues to consider, such as machine derating and torque pulsations caused by distorted currents, interaction between the converter voltage waveform and the motor windings, effects of switching harmonics like reflections and overvoltage, premature bearing failures, and common mode voltages, among others. Some of these technical characteristics can be improved employing multilevel converters due to their capability to generate multistep voltage waveforms and less distorted currents.

1.5   Aim of the Book

The previous discussion opens an exciting perspective around the study of multilevel converters and their use in electric power processing. In this context, the book is intended as a consulting tool for engineers who are interested in state of the art multilevel power converters and their applications. The book is primarily conceived to give the reader a deep insight to multilevel converters, mainly those with a common DC bus. For this, the book presents a first part in which a systematic topological analysis is proposed for the derivation of the main multilevel circuit structures. This methodology is based on the concept of basic circuit cells and switching functions and allows us to derivate the main multilevel topologies. An extensive analysis is presented regarding the topological characteristics of switching circuits, with special emphasis on the topologies with a common DC bus, namely, the diode-clamped multilevel converter, the flying capacitor multilevel converter, and the asymmetric cascaded multilevel converter. For each topology, the particular control issues are exposed and proper modulation and control strategies are thoroughly developed. At the end of the book, two different case studies are presented regarding the applications of the cascade asymmetric multilevel converter and the diode-clamped multilevel converter, in which the associated control problems are addressed through proper control schemes.

References

1. A. Pandey, B. Singh, B.N. Singh, A. Chandra, K. Al-Haddad, D.P. Kothari. A Review of Multilevel Power Converters. *Journal of the Institution of Engineers*, 86, 220, 231, 2006.
2. J. Rodriguez, S. Bernet, B. Wu, J.O. Pontt, S. Kouro. Multilevel Voltage-Source-Converter Topologies for Industrial Medium-Voltage Drives. *IEEE Transactions on Industrial Electronics*, 54(6), 2930–2945, 2007.
3. D. Krug, M. Malinowski, S. Bernet. Design and Comparison of Medium Voltage Multi-Level Converters for Industry Applications. In *IEEE Industry Applications Annual Meeting (IAS'04)*, Seattle, WA, October 3–7, 2004, vol. 2, pp. 781–790.
4. L.G. Franquelo, J. Rodriguez, J.I. Leon, S. Kouro, R. Portillo, M.A.M. Prats. The Age of Multilevel Converters Arrives. *IEEE Industrial Electronics Magazine*, 2(2), 28–39, 2008.
5. H.S. Choi, J.G. Cho, G.H. Cho. A General Circuit Topology of Multilevel Inverter. In *22nd Annual IEEE Power Electronics Specialists Conference (PESC'91)*, Cambridge, MA, June 24–27, 1991, pp. 96–103.
6. B. Wu. *High-Power Converters and AC Drives*. John Wiley & Sons, Hoboken, NJ, 2006.
7. L.M. Tolbert, F.Z. Peng, T.G. Habetler. Multilevel Converters for Large Electric Drives. *IEEE Transactions on Industry Applications*, 35(1), 36–44, 1999.
8. S. Bernet. Recent Developments of High Power Converters for Industry and Traction Applications. *IEEE Transactions on Power Electronics*, 15(6), 1102–1117, 2000.
9. C. Meyer, R.W. De Doncker. Power Electronics for Modern Medium-Voltage Distribution Systems. In *The 4th International Power Electronics and Motion Control Conference (IPEMC'04)*, Xian, China, August 14–16, 2004, pp. 58–66.
10. S. Kouro, M. Malinowski, K. Gopakumar, J. Pou, L.G. Franquelo, B. Wu, J. Rodriguez, M. Pérez, J.I. Leon. Recent Advances and Industrial Applications of Multilevel Converters. *IEEE Transactions on Industrial Electronics*, 57(8), 2553–2580, 2010.
11. H. Abu-Rub, J. Holtz, J. Rodriguez, G. Baoming. Medium-Voltage Multilevel Converters—State of the Art, Challenges, and Requirements in Industrial Applications. *IEEE Transactions on Industrial Electronics*, 57(8), 2581–2596, 2010.
12. J. Rodriguez, L.G. Franquelo, S. Kouro, J.I. Leon, R.C. Portillo, M.A.M. Prats, M.A. Perez. Multilevel Converters: An Enabling Technology for High-Power Applications. *Proceedings of the IEEE*, 97(11), 1786–1817, 2009.
13. A. Ghosh, G. Ledwich. *Power Quality Enhancement Using Custom Power Devices*. Kluwer Academic Publishers, Boston, 2002.
14. R.C. Dugan, M.F. McGranaghan, H.W. Beaty. *Electrical Power Systems Quality*. McGraw-Hill, New York, 1996.
15. N.G. Hingorani, L. Gyugyi. *Understanding FACT, Concepts and Technology of Flexible AC Transmission Systems*. IEEE Press, Piscataway, NJ, 2000.
16. E. Acha, V.G. Agelidis, O. Anaya-Lara, T.J.E. Miller. *Power Electronic Control in Electrical Systems*. Newnes Power Engineering Series, Oxford, UK, 2002.
17. M.M. Morcos, J.C. Gomez. Electric Power Quality—The Strong Connection with Power Electronics. *IEEE Power and Energy Magazine*, 1(5), 18–25, 2003.

18. Y. Yamaguchi, K. Oota, K. Kurachi, F. Tokunoh, H. Yamaguchi. A 6kv/5ka Reverse Conducting GCT. In *IEEE Industry Applications Annual Meeting (IAS'01)*, Chicago, October 1–4, 2001, pp. 1497–1503.

19. A. Weber, P. Kern, T. Dalibor. A Novel 6.5 kV IGCT for High Power Current Source Inverters. *ABB Semiconductors*, June 2001.

20. P.K. Steimer, H.E. Gruning, J. Werninger, E. Carroll, S. Klaka, S. Linder. IGCT—A New Emerging Technology for High Power, Low Cost Inverters. In *IEEE Industry Applications Annual Meeting (IAS'97)*, New Orleans, October 5–9, 1997.

21. T. Setz, M. Lüscher. *Applying IGCTs*. ABB Switzerland Ltd. Semiconductors, Application Note, Doc. 5SYA2032-03, www.abb.com/semiconductors, www.abb.com/powerelectronics, October 2007.

22. P.K. Steimer, O. Apeldoorn, B. Odegard, S. Bernet, T. Brückner. Very High Power IGCT PEBB Technology. *IEEE Power Electronics Specialists Conference (PESC'05)*, Recife, Brazil, June 2005, pp. 1–7.

23. B.J. Baliga. The Future of Power Semiconductor Device Technology. *Proceedings of the IEEE*, 89(6), 822–832, 2001.

24. S. Chen, A. Géza Joós. Transformerless STATCOM Based on Cascaded Multilevel Inverters with Low Switching Frequency Space Vector PWM. In *9th European Conference on Power Electronic and Applications (EPE'01)*, Graz, Austria, August 27–29, 2001.

25. K. Satoh, M. Yamamoto. The Present State of the Art in High-Power Semiconductor Devices. *Proceedings of the IEEE*, 89(6), 813–820, 2001.

26. N. Mohan, T.M. Undeland, W.P. Robbins. *Power Electronics: Converters, Applications and Design*, 2nd ed. John Wiley & Sons, New York, 2002.

27. D.G. Holmes, T.A. Lipo. *Pulse Width Modulation for Power Converters: Principles and Practice*. IEEE Press, Piscataway, NJ, 2003.

28. A. Nabae, I. Takahashi, H. Akagi. A New Neutral-Point-Clamped PWM Inverter. *IEEE Transactions on Industry Applications*, 17(5), 518–523, 1981.

29. T.A. Meynard, H. Foch. Multi-Level Conversion: High Voltage Choppers and Voltage-Source Inverters. In *IEEE Power Electronics Specialists Conference (PESC'92)*, Toledo, Spain, June 29–July 3, 1992, vol. 1, pp. 397–403.

30. S.A. González, M.I. Valla, C.F. Christiansen. Design of a Tuned Balancing Network for Flying Capacitor Multilevel Converters. In *IEEE Power Electronics Specialists Conference (PESC'05)*, Recife, Brazil, June 12–16, 2005, pp. 1046–1051.

31. J.-S. Lai, F.Z. Peng. Multilevel converters—A new breed of power converters. In *IEEE Industry Applications Annual Meeting (IAS'95)*, Orlando, FL, October 8–12, 1995, vol. 3, pp. 2348–2356.

32. F.Z. Peng. A Generalized Multilevel Inverter Topology with Self Voltage Balancing. *IEEE Transactions on Industry Applications*, 37(2), 611–618, 2001.

33. J. Rodriguez, J.-S. Lai, F.Z. Peng. Multilevel Inverters: A Survey of Topologies, Controls, and Applications. *IEEE Transactions on Industrial Electronics*, 49(4), 724–738, 2002.

34. J. Von Bloh, R.W. De Doncker. Design Rules for Diode-Clamped Multilevel Inverters Used in Medium-Voltage Applications. In *VIII IEEE International Power Electronics Congress (CIEP'02)*, Guadalajara, Mexico, October 20–24, 2002, pp. 165–170.

35. M. Winkelnkemper, F. Wildner, P. Steimer. Control of a 6MVA Hybrid Converter for a Permanent Magnet Synchronous Generator for Windpower. In *18th International Conference on Electrical Machines (ICEM 2008)*, Vilamoura, Portugal, September 6–9, 2008, pp. 1–6.

36. K. Fujii, U. Schwarzer, R.W. De Doncker. Comparison of Hard-Switched Multi-Level Inverter Topologies for STATCOM by Loss-Implemented Simulation and Cost Estimation. In *IEEE Power Electronics Specialists Conference (PESC'05)*, Recife, Brazil, June 12–16, 2005, pp. 340–346.

37. Y. Cheng, C. Qian, M.L. Crow, S. Pekarek, S. Atcitty. A Comparison of Diode-Clamped and Cascaded Multilevel Converters for a STATCOM with Energy Storage. *IEEE Transactions on Industrial Electronics*, 53(5), 1512–1521, 2006.

38. D. Soto, T.C. Green. A Comparison of High-Power Converter Topologies for the Implementation of FACTS Controllers. *IEEE Transactions on Industrial Electronics*, 49(5), 1072–1080, 2002.

39. S.A. Khajehoddin, A. Bakhshai, P.K. Jain. A Voltage Balancing Method and Its Stability Boundary for Five-Level Diode-Clamped Multilevel Converters. In *IEEE Power Electronics Specialists Conference (PESC 07)*, Orlando, FL, June 17–21, 2007, pp. 2204–2208.

40. T. Brückner, S. Bernet. Loss Balancing in Three-Level Voltage Source Inverters Applying Active NPC Switches. In *IEEE Power Electronics Specialists Conference (PESC01)*, Vancouver, Canada, June 17–21, 2001, pp. 1135–1140.

41. L. Zhang, S.J. Watkins. Capacitor Voltage Balancing in Multilevel Flying Capacitor Inverters by Rule-Based Switching Pattern Selection. *IET Electric Power Applications*, 1(3), 339–347, 2007.

42. J. Dixon, M. Ortuzar, R. Carmi, P. Barriuso, P. Flores, L. Moran. Static Var Compensator and Active Power Filter with Power Injection Capability, Using 27-Level Inverters and Photovoltaic Cells. *IEEE International Symposium on Industrial Electronics (ISIE'06)*, Montreal, Canada, July 9–13, 2006, pp. 1106–1111.

43. M.D. Manjrekar, P.K. Steimer, T.A. Lipo. Hybrid Multilevel Power Conversion System: A Competitive Solution for High-Power Applications. *IEEE Transactions on Industry Applications*, 36(3), 834–841, 2000.

44. Y.S. Lai, F.S. Shyu. Topology for Hybrid Multilevel Inverter. *IEE Proceedings on Electric Power Applications*, 149(6), 449–458, 2002.

45. D. Kai, Z. Yumping, L. Lei, W. Zhichao, J. Hongyuan, Z. Xudong. Novel Hybrid Cascade Asymmetric Inverter Based on 5-Level Asymmetric Inverter. In *IEEE Power Electronics Specialists Conference (PESC'05)*, Recife, Brazil, June 12–16, 2005, pp. 2302–2306.

46. C. Rech, J.R. Pinheiro. Hybrid Multilevel Converters: Unified Analysis and Design Considerations. *IEEE Transactions on Industrial Electronics*, 54(2), 2031–2036, 2007.

47. B.-R. Lin, T.-Y. Yang. Analysis and Implementation of a Three-Level Active Filter with a Reduced Number of Power Semiconductors. *IEE Proceedings on Electric Power Applications*, 152(5), 1055–1064, 2005.

48. P. Barbosa, P. Steimer, J. Steinke, L. Meysenc, M. Winkelnkemper, N. Celanovic. Active Neutral-Point-Clamped Multilevel Converters. In *IEEE Power Electronics Specialists Conference (PESC'05)*, Recife, Brazil, June 12–16, 2005, pp. 2296–2301.

49. S. Mariethoz, A. Rufer. Design and Control of Asymmetrical Multilevel Inverters. In *IEEE Annual Conference of the Industrial Electronics Society (IECON'02)*, Seville, Spain, November 5–8, 2002, vol. 1, pp. 840–845.

50. S.A. González, M.I. Valla, C.F. Christiansen. Analysis of a Cascade Asymmetric Topology for Multilevel Converters. In *IEEE International Symposium on Industrial Electronics (ISIE 2007)*, Vigo, Spain, June 4–7, 2007, pp. 1027–1032.

51. S.A. González, M.I. Valla, C.F. Christiansen. 5-Level Cascade Asymmetric Multilevel Converter. *IET Power Electronics*, 3(1), 120–128, 2010.

52. A. Cetin, H.F. Bilgin, A. Acik, T. Demirci, K.N. Kose, A. Terciyanli, B. Gultekin, N. Aksoy, B. Mutluer, I. Cadirei, M. Ermis, K. Ongan, N. Akinei. Reactive Power Compensation of Coal Conveyor Belt Drives by Using D-STATCOMs. In *IEEE Industry Applications Annual Meeting (IAS'07)*, New Orleans, September 23–27, 2007, pp. 1731–1740.

53. V. Immanuel, G. Yankanchi. Voltage Sag Compensation Technique for Three-Level Voltage Source Inverter Based Dynamic Voltage Restorer. In *IEEE International Symposium on Industrial Electronics (ISIE'06)*, Montreal, Canada, July 9–13, 2006, pp. 1652–1657.

54. D.M. Vilathgamuwa, H.M. Wijekoon, S.S. Choi. A Novel Technique to Compensate Voltage Sags in Multiline Distribution System: The Interline Dynamic Voltage Restorer. *IEEE Transactions on Industrial Electronics*, 53(5), 1603–1611, 2006.

55. A.E. León, M.F. Farías, P.E. Battaiotto, J.A. Solsona, M.I. Valla. Control Strategy of a DVR to Improve Stability in Wind Farms Using Squirrel-Cage Induction Generators. *IEEE Transactions on Power Systems*, 26(3), 1609–1617, 2011.

56. M. Basu, S.P. Das, G.K. Dubey. Performance Study of UPQC-Q for Load Compensation and Voltage Sag Mitigation. In *IEEE Annual Conference of the Industrial Electronics Society (IECON'02)*, Seville, Spain, November 5–8, 2002, vol. 1, pp. 698–703.

57. L.M. Tolbert, T.J. King, B. Ozpineci, J.B. Campbell, G. Muralidharan, D.T. Rizy, A.S. Sabau, H. Zhang, W. Zhang, Y. Xu, H.F. Huq, H. Liu. Power Electronics for Distributed Energy Systems and Transmission and Distribution Applications. Report prepared by the Oak Ridge National Laboratory for the U.S. Department of Energy, code ORNL/TM-2005/230, 2005.

58. S.A. Verne, M.I. Valla. Direct Connection of WECS System to the MV Grid with Multilevel Converters. *Renewable Energy*, 41, 336–344, 2012.

59. L.H. Hansen, L. Helle, F. Blaabjerg, E. Ritchie, S. Munk-Nielsen, H. Bindner, P. Sörensen, B. Bak-Jensen. Conceptual Survey of Generators and Power Electronics for Wind Turbines. Report prepared by the Riso National Laboratory, Roskilde, Denmark, code Riso R-1205(EN), 2001.

60. Z. Chen, J. Guerrero, F. Blaajberg. A Review of the State of the Art of Power Electronics for Wind Turbines. *IEEE Transactions on Power Electronics*, 24(8), 1859–1875, 2009.

61. M. Tsili, S. Papathanassiou. A Review of Grid Code Technical Requirements for Wind Farms. *IET Renewable Power Generation*, 3(3), 308–332, 2009.

62. P. Rodriguez, A. Timbus, R. Teodorescu, M. Liserre, F. Blaabjerg. Reactive Power Control for Improving Wind Turbine System Behavior Under Grid Faults. *IEEE Transactions on Power Electronics*, 24(7), 1798–1801, 2009.

63. M. Dahlgren, H. Frank, M. Leijon, F. Owman, L. Walfridsson. Windformer: Energía Eólica a Gran Escala. *Revista ABB*, 3, 31–37, 2000.
64. J.M. Carrasco, L.G. Franquelo, J.T. Bialasiewicz, E. Galvan, R.C.P. Guisado, M.M. Prats, J.I. Leon, N. Moreno-Alfonso. Power Electronic Systems for the Grid Integration of Renewable Energy Sources: A Survey. *IEEE Transactions on Industrial Electronics*, 53(4), 1002–1016, 2006.
65. R.A. Hanna, S.W. Randall. Medium-Voltage Adjustable-Speed-Drive Retrofit of an Existing Eddy-Current Clutch Extruder Application. *IEEE Transactions on Industry Applications*, 36(6), 1750–1755, 2000.

chapter 2

Multilevel Topologies

2.1 Introduction

Voltage source multilevel converters (VSMCs) have been developed in the last years for high-power applications, due to their capability to work at medium-voltage levels without power transformers. The blocking voltage of the power devices has limited the supply voltage of transformer less power converters. But this limit can be overcome if the number of voltage levels of the converter is increased. Moreover, the increment of voltage levels contributes to build a softer AC voltage with the consequent reduction in harmonic distortion. Also, the voltage variation rate (dv/dt) is reduced, diminishing the electromagnetic interference (EMI) problems and other stresses on the power switches.

The classic multilevel converters, like the diode-clamped (DCMC), flying capacitor (FCMC), or cascade H-bridge (CCMC), present modular structures. So in all of them it is simple to increase the voltage levels adding basic modules. The different combinations of ON and OFF states of the power switches of the VSMC determine the "switching states" of the converter. Each state defines one level of the output voltage. Depending on the topology, a given voltage level may be built with different switching states. These states are known as the redundant states of the converter. In some topologies, like the DCMC, there exist some states for which the voltage level is not determined by the converter, but depends on the sign of the load current. These states are defined as forbidden states and should be avoided. In general, the multilevel topologies have more switching states than the number of voltage levels that can be synthesized.

It is possible to think of an ideal multilevel topology as that which defines one voltage level for each state combination of power switches; so this ideal converter has neither redundant states nor forbidden states. But this is not the case for actual topologies. The redundant and forbidden states increase much more than the number of voltage levels, with the consequent increment in complexity, volume, and cost of the multilevel converters.

A systematic analysis of the classical multilevel topologies is developed in this chapter. The aim of this analysis is to give the basis to the development of new topologies with more voltage levels and reduced complexity. The multilevel topologies are divided into two main groups; those that use a common DC source and those that require multiple isolated DC sources. Moreover, they can be divided between symmetric, asymmetric, and hybrid topologies. The systematic analysis, together with the introduction of a new method with a graphic representation of the switching states and the voltage levels, allows visualizing and understanding the behavior of multilevel converters.

The chapter ends with the presentation of a new five-level hybrid topology that comprises the cascade connection of two different topologies. The ratio between the available voltage levels and the switching states is greatly improved when compared to the five-level classic topologies: DCMC, FCMC, and CCMC.

2.2 Generalized Topology with a Common DC Bus

Figure 2.1 shows one leg of a symmetric *n*-level VSMC with a common DC bus. This is built with a specific array of power switching devices and

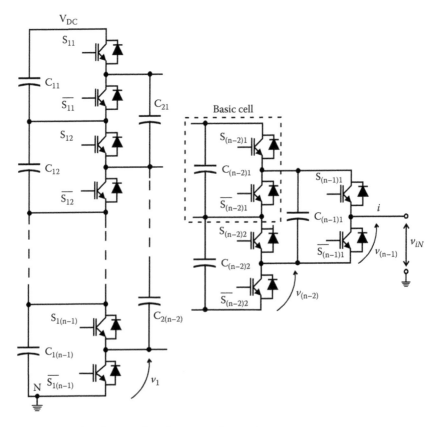

Figure 2.1 N-level generalized symmetric topology.

capacitors. It is named the generalized topology introduced by Peng in 2001 [1]. It has a cellular structure in which the number of voltage levels is increased by adding basic cells. The basic cell is the functional unit of the generalized topology, and it is built with two complementary power switches and one capacitor. An n-level converter has $n - 1$ stages. Each stage is formed by a stack of basic cells. This cellular structure allows us to increase the number of voltage levels by implementing a symmetrical growth, in both vertical and horizontal directions.

2.2.1 Basic Cell

The basic cell, shown in Figure 2.2, is formed with two complementary power switches $(S_j, \bar{S}_j)$, which can carry bipolar current, together with a power source (battery, solar panel, etc.) or a power storage element (capacitor).

The complementary power switches are controlled with a switching function s_j, which has two possible states,

$$s_j = \begin{cases} 1 \\ 0 \end{cases} \qquad (2.1)$$

When the switching function equals 1, S_j is ON while $\bar{S}_j$ is OFF. On the other hand, when the switching function equals 0, S_j is OFF while $\bar{S}_j$ is ON. The output voltage of the cell is measured between nodes g and $g-$, and it is defined by the switching function. Assuming that the voltage

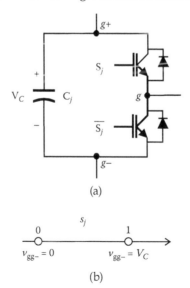

(a)

(b)

Figure 2.2 Basic cell: (a) topology and (b) graphic representation of cell states.

variation on C_j is negligible compared to its average value V_C, the output voltage results:

$$v_{(g,g-)} = s_j \cdot V_C \qquad (2.2)$$

The basic cell has two voltage levels determined by the switching function. State 0 defines a 0 V output voltage and state 1 defines an output voltage equal to V_C. Figure 2.2b shows a graphic representation of the switching states of the cell, together with the corresponding output voltage. This way of representing the switching states of the converter allows obtaining a compact representation of the converter behavior in more complex topologies.

The basic cell is the most elementary topology of a voltage source converter. It corresponds with one leg of a two-level topology. In this way, the classic two-level voltage source inverter is included in the generalized topology with the minimum number of voltage levels.

2.2.2   Generalized Topology Characteristics

As shown in Figure 2.1, one leg of an n-level generalized topology is formed by $(n-1)$ cascaded stages. Each stage employs one less cell than the previous one. So, the first stage, near the DC source, has $(n-1)$ vertically connected basic cells, while the last one, near the load, has only one.

Every basic cell should have the same voltage (V_C) in order to synthesize a symmetric output voltage. The voltage level V_C is fixed by the capacitor voltage divider of the first stage. Assuming that all the capacitors have the same capacitance, then V_C is

$$V_C = \frac{V_{DC}}{n-1} \qquad (2.3)$$

A constant and equal voltage source for each cell V_C in every stage is guaranteed when all the cells in one stage use the same switching function. Figure 2.3 shows the mechanism to equalize the voltages in each cell of every stage. The figure represents a portion of the first three stages of an n-level leg. The power switches that are OFF are indicated in gray, while those that are ON are drawn in black. It is clearly seen that the switches S_{1j} connect the capacitors C_{2j} in parallel with C_{1j} (where $j = 1, 2, \ldots$ $n-1$). At the same time, capacitors C_{3j} are connected in parallel with $C_{2(j+1)}$ through the switches $\bar{S}_{2j}$. When switches S_{1j} are turned OFF, the first stage changes its state and the capacitors C_{2j} are now connected in parallel with $C_{1(j+1)}$. A similar change can be observed in the other stages of the leg. In

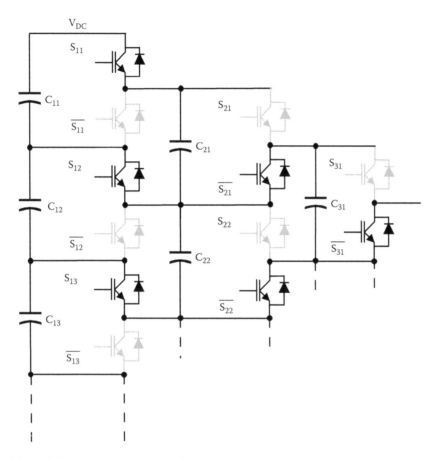

Figure 2.3 Parallel connection of basic cells.

this way the voltage V_C is maintained in all the cells of the generalized topology. It is important to remark that the voltage on every capacitor is preserved when all the cells in the same stage commutate simultaneously with the same switching function.

The output voltage of the leg (v_{iN} in Figure 2.1) is defined by the sum of the voltages of all the cells that are below the output connection. Taking into account (2.2), the output voltage can be written as

$$v_{iN} = V_C \cdot \sum_{l=1}^{n-1} s_l \qquad (2.4)$$

where s_l is the switching function of the l stage.

All the cells in each stage are controlled with the same switching function, and it defines two possible states. Then, an n-level topology with $(n - 1)$ stages has $2^{(n-1)}$ possible states. It is possible to imagine an $n - 1$ dimensional space in which the switching function of each stage is represented in each axis of this space. This graphic representation allows summarizing the characteristics of the topology and its voltage levels. Moreover, it becomes a useful tool to select the proper switching strategy to control the voltage balance on all the capacitors of any VSMC [2].

2.2.3   Three-Level Generalized Topology

The three-level generalized topology has three basic cells in each leg, forming two stages, as can be seen in Figure 2.4a. The first one has two cells and is controlled by a switching function s_1. The other has only one cell and is controlled by a switching function s_2. The output voltage of each leg results in:

$$v_{iN} = V_C \cdot \left(s_1 + s_2\right) \tag{2.5}$$

The graphic representation of the states and voltages for the three-level topology is shown in Figure 2.4b. Each switching function is located on each axis, so the corners of the square represent each state and voltage level.

Figure 2.5 shows the four connections or circuits resulting for each state of the converter. The switches that are OFF are indicated in light gray. State 00 occurs when the switches $\bar{S}_{1(1,2)}$ and $\bar{S}_2$ are ON (Figure 2.5a). Then, the output voltage equals 0 V and the capacitor C_2 is connected in parallel with C_{12}. When the second stage changes and S_2 is turned on, state 01 takes place (Figure 2.5b). The capacitor C_2 is still in parallel with C_{12}, and they apply the voltage across them to the output, which equals $V_{DC}/2$. For state 11 (Figure 2.5c) the switches $S_{1(1,2)}$ and S_2 are ON and the output voltage equals V_{DC}. In this case, the capacitor C_2 is connected in parallel with C_{11}. Finally, for state 10 (Figure 2.5d) the output voltage is defined by the voltage over capacitor C_{12} and equals $V_{DC}/2$. It is clear that there are two states, 01 and 10, that generate the same output voltage. So, these are defined as redundant states. The connection of capacitors is different in both states, so a careful usage of the redundant states may serve to correct possible unbalances of the voltages on the capacitors. In order to reduce the total number of switching transitions, it is convenient to change only one switching function at a time. This defines the adjacent states, which are those that differ in only one switching function. Minimum voltage steps are guaranteed on the output voltage when the converter jumps between adjacent states.

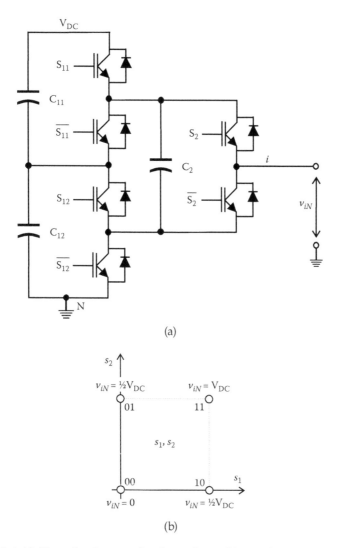

(a)

(b)

Figure 2.4 (a) Three-level generalized topology. (b) Graphic representation of converter states and voltage levels.

2.3   Converters Derived from the Generalized Topology

The generalized topology allows increasing the number of voltage levels by simply adding new basic cells. For example, to increase the voltage levels from n to $n + 1$ it is necessary to add a new stage with n cells. It is an easy way to increase the voltage levels, but it requires a lot of devices. Then, the implementation complexity of the topology also increases. Moreover,

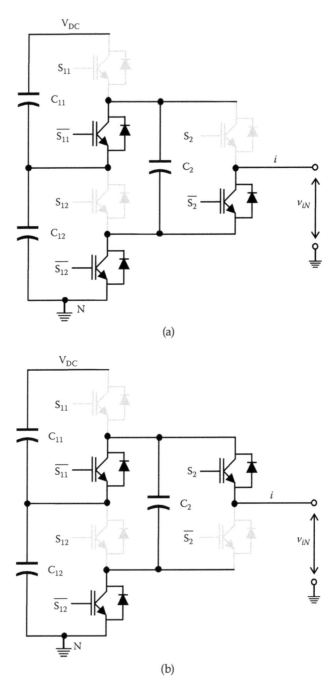

Figure 2.5 Converter states of a three-level generalized topology: (a) state 00, (b) state 01, (c) state 11, and (d) state 10.

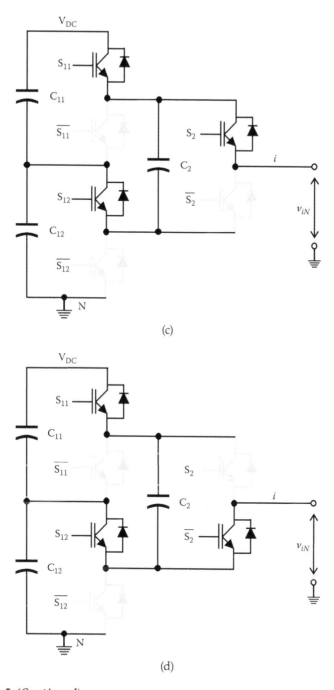

(c)

(d)

Figure 2.5 (Continued)

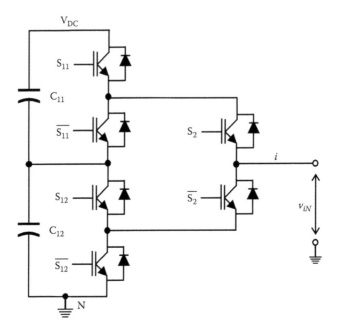

Figure 2.6 ANPC topology.

the number of possible states increases more than the voltage levels, also increasing the number of redundant states. So, the generalized topology is useful to understand the operation of multilevel topologies, but its practical implementation is not convenient when the number of levels increases.

Classical topologies like the diode-clamped multilevel converter and the flying capacitor multilevel converter may be seen as simplifications of the generalized topology, which preserve its symmetry.

2.3.1 Diode-Clamped Topology

The previous analysis of the three-level generalized topology shows that C_2 is always connected in parallel with one of the capacitors of the first stage. Then, it is possible to eliminate C_2 without altering the voltage levels at the output. Figure 2.6 shows the converter without C_2, which is called active neutral point clamping (ANPC) [3,4].

A further reduction in semiconductor devices may be obtained by eliminating the transistors from the switches S_{12} and $\bar{S}_{11}$ and leaving only the diodes (D_e and $\bar{D}_e$), as shown in Figure 2.7. In this case the voltages synthesized by states 00 and 11 are the same as before. But the voltage generated in states 01 and 10 depends on the sign of the current in node i. Figure 2.7 shows the connections in these two states, where the switches being OFF are indicated in light gray. Analyzing state 01 (Figure 2.7a),

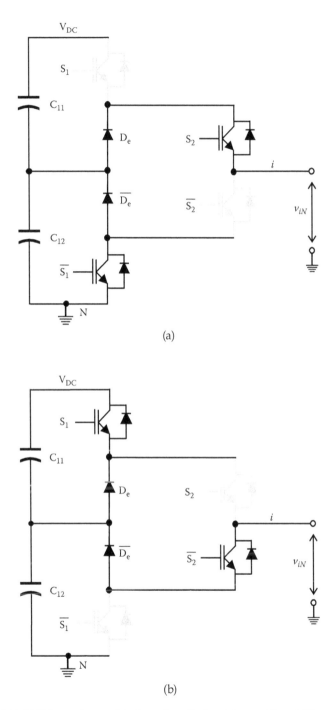

Figure 2.7 ANPC topology with clamping diodes: (a) state 01 and (b) state 10.

when the current enters into node i, it finds its way through the diodes of switches S_2 and S_1. So the resulting voltage v_{iN} equals V_{DC}. On the contrary, when the current goes out of node i, it finds its way through the diodes of switches $\bar{S}_2$ and $\bar{S}_1$ and the resulting voltage is 0. In the case of state 10 (Figure 2.7b), when the current enters in node i, it finds its way through switch $\bar{S}_2$ and diode $\bar{D}_e$, resulting in $v_{iN} = V_{DC}/2$. On the contrary, when the current goes out of node i, it finds its way through the diodes of switches $\bar{S}_2$ and $\bar{S}_1$, resulting in $v_{iN} = 0$. So this topology cannot generate the intermediate-voltage levels and states 01 and 10, which are considered forbidden states. This topology is not a practical multilevel converter.

2.3.1.1   NPC Converter Topology

It is possible to make a small change in the topology exchanging the positions of the switches $\bar{S}_1$ and $\bar{S}_2$, as shown in Figure 2.8. In this way the neutral point clamped (NPC) multilevel converter is obtained [5]. The NPC converter topology has the same states as the three-level generalized topology, whose graphic representation is presented in Figure 2.4b. While states 00 and 11 synthesize the same voltage levels 0 V and V_{DC}, respectively, there exist some differences for the intermediate states 01 and 10. Figure 2.8 shows the intermediate states for the NPC converter. Analyzing state 01 (Figure 2.8a), when the current enters in node i, it finds its way through the switch $\bar{S}_1$ and the diode $\bar{D}_e$, resulting in $v_{iN} = V_{DC}/2$. When the current goes out of node i, it finds its way through the switch S_2 and the diode D_e, resulting in $v_{iN} = V_{DC}/2$. In consequence, state 01 clearly synthesizes the voltage $V_{DC}/2$ independently of the sign of the current. The analysis of state 10 (Figure 2.8b) indicates that when the current enters in node i, it finds its way through the diodes of switches S_2 and S_1, resulting in $v_{iN} = V_{DC}$. On the contrary, when the current goes out of node i, it finds its way through the diodes of switches $\bar{S}_2$ and $\bar{S}_1$, resulting in $v_{iN} = 0$. In consequence, this state is forbidden to avoid any uncertainty in the generation of the voltage levels. Nevertheless, this topology is valid and capable of generating three voltage levels.

It is important to see that the blocking voltage of every switch is fixed at the value $V_{DC}/2$ through diode D_e or $\bar{D}_e$. During state 01 the open switches S_1 and $\bar{S}_2$ are in parallel with capacitors C_{11} and C_{12}, respectively. So the voltage across them is fixed to $V_{DC}/2$ by both capacitors. In state 00 the switches $\bar{S}_1$ and $\bar{S}_2$ are ON, and they block $\bar{D}_e$, but D_e may conduct current connecting S_2 in parallel with C_{12}, clamping its voltage to $V_{DC}/2$.

2.3.1.2   Four-Level Diode-Clamped Topology

Also, the four-level diode-clamped multilevel converter (4L-DCMC) can be obtained from the generalized topology. Starting with a three-stage generalized topology, first all the switches of the internal cells are implemented with diodes (D_{ej} and $\bar{D}_{ej}$, with $j = 1, 2$, and 3). Only the upper and

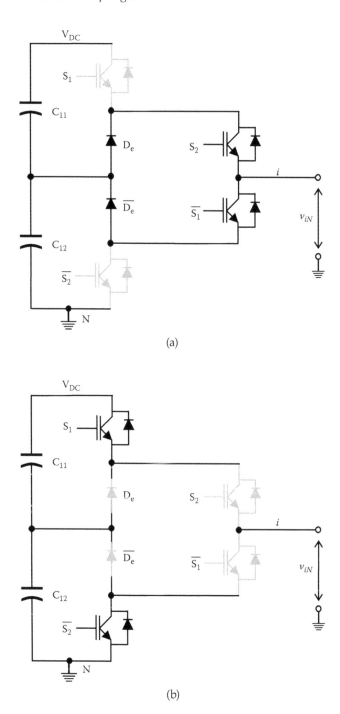

Figure 2.8 NPC converter states: (a) state 01 and (b) state 10.

lower switches of each stage are implemented with transistors and free-wheeling diodes. In a second step the lower switches exchange positions among the stages, as shown in Figure 2.9a.

Figure 2.9b shows the graphical representation of the different states of a 4L-DCMC. The four-level generalized topology has three stages and uses three switching functions (s_1, s_2, s_3), which determine the existence of $2^3 = 8$ possible states. In Figure 2.9b the switching functions s_1, s_2, and s_3 are represented along the three axes. They generate a cube whose corners are the different states of the converter. The corresponding voltage levels of v_{iN} are obtained from (2.4) and are also indicated in the figure.

$$v_{iN} = V_C \cdot \left(s_1 + s_2 + s_3 \right) \tag{2.6}$$

with $V_C = V_{DC}/3$.

Figure 2.10 shows the connections for the allowable states of the four-level DCMC, namely, 000, 001, 011, and 111. The devices that are ON are indicated in black. The blocking voltage for each transistor as well as each diode is $V_{DC}/3$. Analyzing state 000, the switches $S_{(1,2,3)}$ are OFF and their complements, which are ON, connect node i to the negative of the DC bus. Then, the clamping diodes $\bar{D}e_j$ (light gray) are blocked, while diodes D_{ej} may potentially conduct some current. In this condition the diodes connect the open switches in parallel with one capacitor of the DC bus. This is shown in Figure 2.10a, where the switches S_1, S_2, and S_3 are in parallel with C_{11}, C_{12}, and C_{13}, respectively. This condition is reproduced for the rest of the allowable states, fixing the blocking voltage of every switch to $V_{DC}/3$.

The same as the NPC converter, the 4L-DCMC has forbidden states for which the output voltage depends on the sign of the output current. They are all represented in Figure 2.11. Starting with state 110 the switches $\bar{S}_1$, $\bar{S}_2$, and S_3 are open (Figure 2.11d). When the load current goes out of node i, it finds its way through the diodes in parallel with switches $\bar{S}_1$, $\bar{S}_2$, and $\bar{S}_3$, resulting in $v_{iN} = 0$ V. On the contrary, when the current goes into node i, it finds its way through the diodes of switches S_1, S_2, and S_3, resulting in $v_{iN} = V_{DC}$. For state 100 (Figure 2.11b) the switches $\bar{S}_1$, S_2, and S_3 are open, and a detailed analysis shows that ingoing and outgoing currents again generate $v_{iN} = V_{DC}$ and $v_{iN} = 0$ V. For state 101 (Figure 2.11c) the switches $\bar{S}_1$, S_2, and $\bar{S}_3$ are open, and again an ingoing current forces $v_{iN} = V_{DC}$ at node i. On the contrary, an outgoing current finds its way through S_3 and the clamping diodes De_2 and De_3, resulting in $v_{iN} = V_{DC}/3$. Finally, for state 010 (Figure 2.11a), S_1, $\bar{S}_2$, and S_3 are open, and an outgoing current forces $v_{iN} = 0$ V through the diodes of the complementary switches. On the other hand, an ingoing current in node i finds its way through $\bar{S}_1$ and the diodes De_1 and $\bar{D}e_2$, resulting in $v_{iN} = 2V_{DC}/3$.

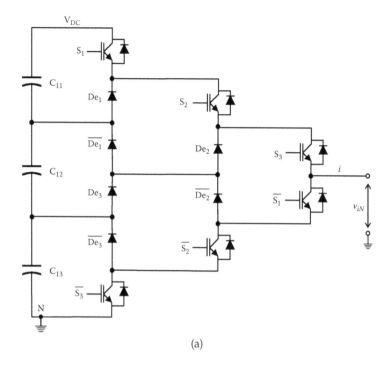

(a)

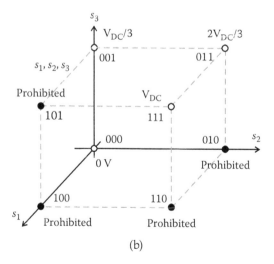

(b)

Figure 2.9 (a) Four-level DCMC. (b) State diagram and voltages.

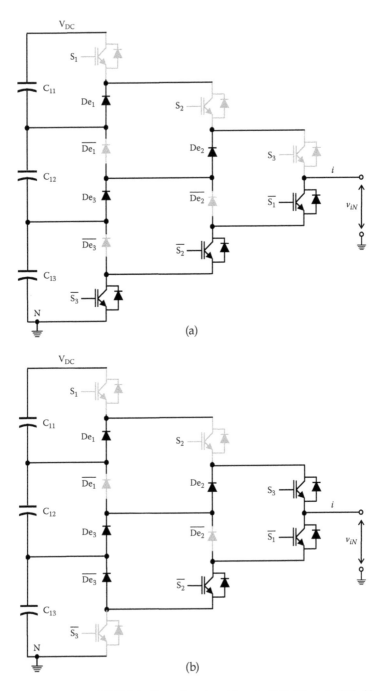

Figure 2.10 Four-level DCMC allowed states: (a) state 000, (b) state 001, (c) state 011, and (d) state 111.

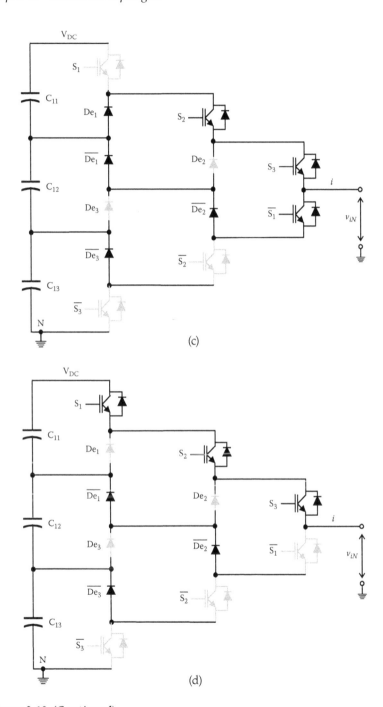

Figure 2.10 (Continued)

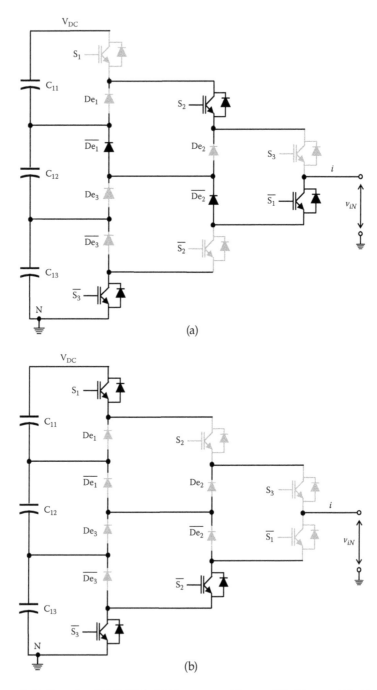

Figure 2.11 Four-level DCMC forbidden states: (a) state 010, (b) state 100, (c) state 101, and (d) state 110.

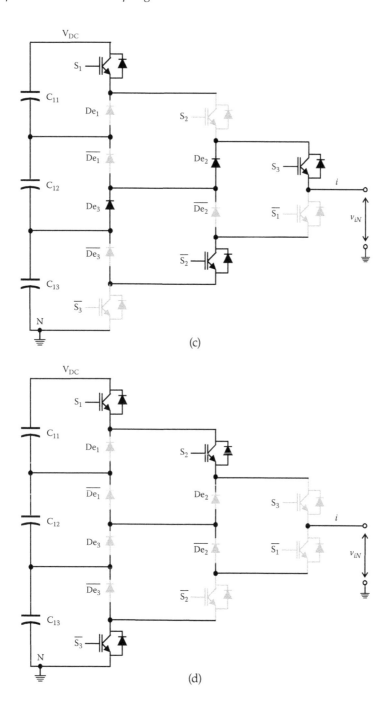

Figure 2.11 (Continued)

It is possible to obtain some preliminary conclusions about allowable and forbidden states for an n-level DCMC. Figure 2.10 shows that for each allowable state there exists two antiparallel ways to connect node i with the same point of the DC bus. In this way, the current going in or out node i always connects the output to the same voltage. On the other hand, these two ways cannot be found for the forbidden states shown in Figure 2.11. In general, a forbidden state is present when, at least one switch of the high index is OFF, while one or more switches of the lower index are ON. In other words, for an allowable state, all the closed switches must be contiguous. If there exist one open switch between two closed switches, that is a forbidden state. This means that there is only one allowable state for each voltage level.

2.3.1.3   Five-Level Diode-Clamped Topology

Figure 2.12a shows a five-level diode-clamped multilevel topology. It is obtained from a four-stage generalized topology, following the same steps to reduce components as shown for the 4L-DCMC. In this case there are four switching functions defining 16 different states.

Figure 2.12b shows the graphic representation of the topology states. The four switching functions, s_1, s_2, s_3, and s_4, are represented in two cubes, one related to $s_4 = 0$ and the other to $s_4 = 1$. The corners of each cube represent all the possible combinations and the voltage levels in each allowable state. Both cubes are aligned along their diagonals, which contain states 0000, 0001, 1110, and 1111. The adjacent states are defined by the contiguous corners of one cube as well as the neighbor corners of both cubes. The number of allowable states equals the number of synthesized voltage levels. So there are no redundant states in each leg, and there are 11 forbidden states. The allowable states to reach every voltage level from 0 to V_{DC} (states 0000 to 1111) are indicated with white circles in Figure 2.12b.

The voltage over each capacitor of the DC link is $V_{DC}/4$, and this is the blocking voltage required to each power switch.

2.3.1.4   n-Level DCMC Topologies

The diode-clamped multilevel topologies preserve the number of stages of the generalized topology. They also preserve the connection of the internal nodes of the contiguous stages. The diodes De_j and $\bar{D}e_j$ conduct in a complementary way for the intermediate voltages. When the current goes out of node i, the diodes De_j of each stage conduct, while when the current goes into node i, the complementary diodes $\bar{D}e_j$ of each stage conduct. For example, considering the 4L-DCMC and state 011 shown in Figure 2.10c, the active diodes are De_1, $\bar{D}e_1$, and $\bar{D}e_2$. Diodes $\bar{D}e_1$ and $\bar{D}e_2$ connected in series carry an ingoing load current. For an outgoing current, only De_1 will conduct the current. Another example, taking the 5L-DCMC shown in Figure 2.12a, for state 0111 the diodes $\bar{D}e_1$, $\bar{D}e_4$, and

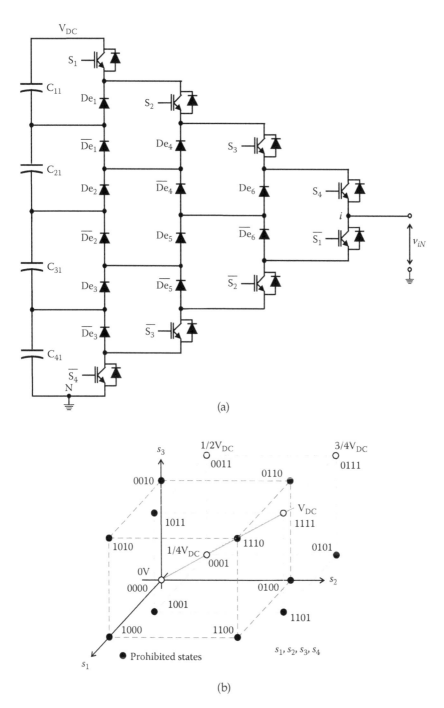

Figure 2.12 Five-level DCMC: (a) topology and (b) state diagram and voltages.

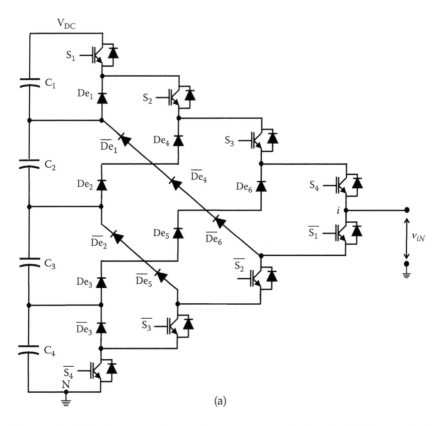

Figure 2.13 Minimum number of components: five-level DCMC practical implementation.

$\bar{De}_6$ are connected in series with $\bar{S}_1$ to carry an ingoing current, while De_1 is connected in series with S_2, S_3, and S_4 to conduct an outgoing current. In the case of state 0011, the diodes $\bar{De}_2$ and $\bar{De}_5$ are connected in series with $\bar{S}_1$ and $\bar{S}_2$ and carry the ingoing current, while De_2 and De_4 in series with S_3 and S_4 carry the outgoing current. The previous analysis shows that there are some connections between stages that can be avoided without modifying the operation of the topology. Figure 2.13a shows a new connection of the clamping diodes. If it is accepted that each clamping diode may block different voltages, the number of devices may be further reduced, resulting in the practical implementation shown in Figure 2.13b. In this practical implementation, each stage has a pair of complementary diodes ($De_{(1,2,3)}$ and $\bar{De}_{(1,2,3)}$) with different blocking voltage. For example, De_3 should block $V_{DC}/4$, while $\bar{De}_3$ should block $3V_{DC}/4$. The number of clamping diodes is smaller than the number of active switches. An n-level DCMC requires $2(n-1)$ active switches and $2(n-2)$ clamping diodes [6].

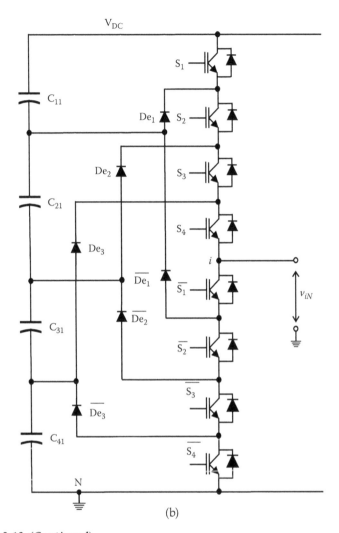

(b)

Figure 2.13 (Continued)

In summary, the diode-clamped multilevel topology preserves the structure and number of stages of the generalized topology. Each stage of a DCMC uses different switching functions because the complementary switches are connected in different stages. The DCMC allows reducing the number of active switches and capacitors, at the expense of having forbidden states that should be avoided. The number of forbidden states equals the number of possible states (2^{n-1}) minus the number of voltage levels (n). Each leg of an n-level topology requires $2(n-1)$ active switches, $(n-1)(n-2)$ clamping diodes with equal blocking voltage, or $2(n-2)$ with different blocking capability. It also requires $(n-1)$ capacitors in the DC link, which

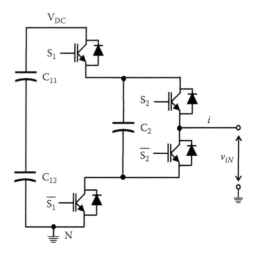

Figure 2.14 Three-level FCMC topology.

are common to all legs of the topology. A DCMC with more than four or five levels is not practical due to the increased complexity, cost, and volume.

2.3.2 Flying Capacitor Topology

While performing the analysis of the three-level generalized topology shown in Figure 2.4a, it was demonstrated that the power switches S_{12} and $\bar{S}_{11}$ connect C_2 in parallel with C_{11} or C_{12}, fixing the average voltage across C_2 to $V_{DC}/2$. It is possible to find a switching sequence between the adjacent states shown in Figure 2.4b, in such a way to maintain the voltage across C_2 equal to $V_{DC}/2$, without using S_{12} and $\bar{S}_{11}$. When both switches are eliminated, C_2 flies between both stages, as shown in Figure 2.14. This topology is known as capacitor-clamped multilevel topology or flying capacitor multilevel converter (FCMC) [7].

The three-level FCMC comes from a two-stage generalized topology, so it has two switching functions and four possible states, the same as were shown in Figure 2.4b. The same as in the previous analysis, states 00 and 11 generate levels 0 V and V_{DC}, respectively. The connections for the intermediate states are shown in Figure 2.15, where the switches that are open are indicated in light gray. Assuming that the voltage over C_2 remains constant, the output voltage for both states equals $V_{DC}/2$, while in state 01 switches $\bar{S}_1$ and S_2 are closed and the voltage v_{iN} is the voltage directly across from C_2. In state 10 the switches S_1 and $\bar{S}_2$ connect C_2 in series with the DC power supply V_{DC}; then v_{iN} equals the difference between the voltages of the DC link and C_2. Moreover, the blocking voltage for each power switch is the voltage on each capacitor, which is $V_{DC}/2$. Unlike the NPC

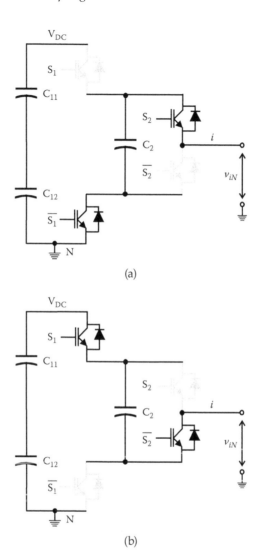

(a)

(b)

Figure 2.15 Three-level FCMC, different states of the topology: (a) state 01 and (b) state 10.

converter, there are no forbidden states and there are two redundant states that generate the same voltage at the output.

2.3.2.1 Voltage on the Flying Capacitor

In the previous section it was assumed that the voltage across C_2 is constant and equal to $V_{DC}/2$. In order to guarantee this constant value, it is necessary to have no net charge on the capacitor along one switching cycle. This condition can be met using the redundant states.

Figure 2.15 helps us to understand the changes of the electric charge on C_2 when the redundant states are applied. It is assumed that the current goes out of node i, and it is almost constant during the switching cycle. For state 01 the current flows through the switches $\bar{S}_1$, S_2, and the capacitor C_2. So, the capacitor loses charge and its voltage decreases. On the other hand, for state 10 the current flows through the DC link, S_1, $\bar{S}_2$, and C_2. Now, the current changes the flowing direction across C_2, so the electric charge of the capacitor increases and its voltage increases. Hence, both states have a complementary action, provided that they are applied for the same amount of time in each switching cycle. So, this condition is required to maintain the voltage level on C_2.

2.3.2.2 *Four-Level Flying Capacitor Topology*

A systematic approach to obtain an n-level FCMC from the generalized topology consists in eliminating the switches from the internal cells while preserving only their capacitors. Figure 2.16a shows the result for a four-level FCMC, where the average voltage on each capacitor equals $V_{DC}/3$. It presents a three-stage structure, where each stage is formed by two complementary switches and one or more flying capacitors. Similarly, as in the generalized topology, in the FCMC each stage is controlled by only one switching function. The 4L-FCMC has three switching functions and eight possible states, as shown in Figure 2.16b. The same as for the 4L-DCMC, the switching functions s_1, s_2, and s_3 define a cube whose corners represent the topology states. The voltage levels calculated with (2.4) are also indicated in each corner. The blocking voltage for each open switch equals the voltage on each capacitor, which is $V_{DC}/3$.

Figure 2.16b shows that there are three redundant states for $V_{DC}/3$ and three redundant states for $2V_{DC}/3$. These states are used to preserve the voltage on the capacitors of each stage, similarly as it was analyzed for the 3L-FCMC. For example, assuming an outgoing constant current in node i (Figure 2.16a), consider the sequence of states 001, 010, and 100. The changes among states are carried out between adjacent states; this means that they need to pass through state 000. Beginning in state 001, the switches $\bar{S}_1$, $\bar{S}_2$, and S_3 are closed and the current flows through C_3, reducing the charge. Afterwards, in state 010, the current flows through $\bar{S}_1$, S_2, $\bar{S}_3$, C_2, and C_3. In this case C_2 loses charge and C_3 recovers the charge lost in state 001. In state 100, the current flows through S_1, $\bar{S}_2$, $\bar{S}_3$, and C_2. In this case C_2 recovers the charge lost in state 010. Then, if the current is almost constant and the topology remains in the different states for the same amount of time, the net charge and the voltage on the flying capacitors are preserved.

2.3.2.3 *Five-Level FCMC*

The five-level FCMC is obtained from the 4L-FCMC, adding one stage with four capacitors, as shown in Figure 2.17a. The voltage on each flying

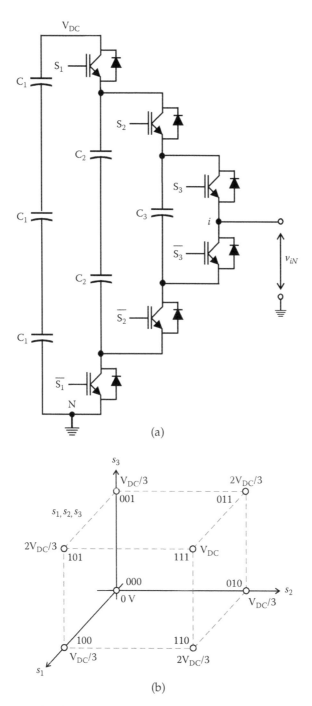

Figure 2.16 Four-level FCMC: (a) topology and (b) state diagram and voltages.

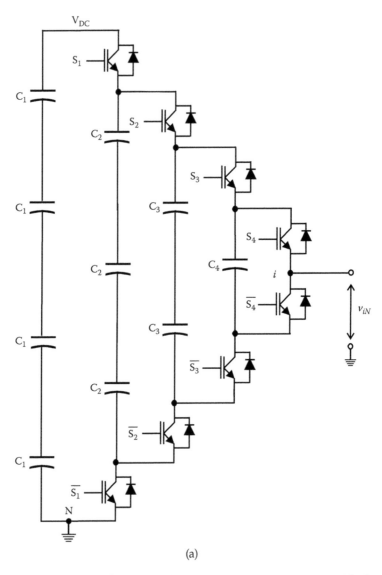

(a)

Figure 2.17 (a) Five-level FCMC. (b) Sixteen states representation with the corresponding voltages.

capacitor is $V_{DC}/4$, which corresponds to the minimum step on the output voltage v_{iN} and the blocking voltage of all power switches. This topology needs four switching functions that generate 16 states, as shown in Figure 2.17b. The graphic representation for a five-level symmetric topology with a common source is done with two cubes, one related to $s_4 = 0$

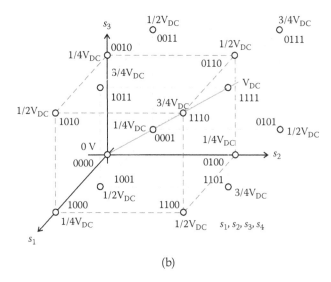

(b)

Figure 2.17 (Continued)

and the other to $s_4 = 1$. The corners of each cube represent all the possible combinations of the switching functions s_1, s_2, and s_3. Both cubes are aligned along their diagonals, which contain states 0000 and 1110 for $s_4 = 0$ and states 0001 and 1111 for $s_4 = 1$. The adjacent states are defined by the contiguous corners of one cube as well as the neighbor corners of both cubes. The 5L-FCMC presents four redundant states for $V_{DC}/4$ and $3V_{DC}/4$ and six redundant states for $V_{DC}/2$. The voltage levels indicated in Figure 2.17b were obtained with (2.4).

It is worth noting that for the FCMC the capacitors on the DC link are common to every leg of the topology, while the flying capacitors belong to each leg independently of the others. Then, an n-level FCMC requires $(n-1)$ capacitors on the DC link plus $(n-1)(n-2)/2$ flying capacitors for each leg. All these capacitors hold the same voltage; this number can be reduced to one capacitor in the DC link plus $(n-2)$ capacitors per leg of the converter if the capacitors support different voltages. The number of active power switches is equal to $2(n-1)$ for each leg, the same as the n-level DCMC and no clamping diodes are required in this topology.

2.4 Symmetric Topologies without a Common DC Link

Figure 2.18 shows one leg of an n-level multilevel topology with independent and isolated power sources. This topology does not have a common DC link for all the legs of the topology. Each stage has two basic cells

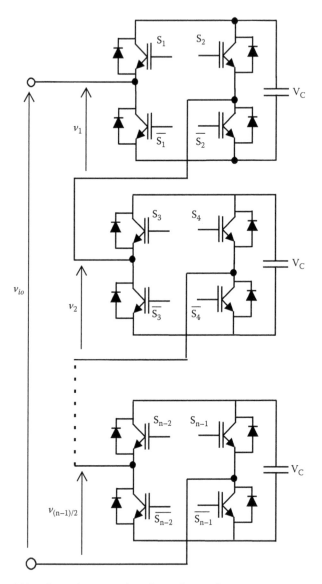

Figure 2.18 N-level topologies with independent voltage sources.

connected in parallel and sharing a common DC source or capacitor. The different stages are connected in series to obtain an n-level voltage at the output. This topology is known as n-level cascaded cell multilevel converter (CCMC) [8,9]. The voltage of each stage is calculated as the voltage difference of each basic cell, as indicated in (2.2). Then, each stage is controlled with two independent switching functions and the stage voltage (v_j) is calculated as

$$v_j = V_C \cdot \left(s_{(2j-1)} - s_{2j} \right) \tag{2.7}$$

where $s_{(2j-1)}$ and $s_{(2j)}$ are the switching functions and V_C is the DC voltage source of the jth stage.

Each stage offers three voltage levels, 0 and $\pm V_C$; a series connection of two stages gives five levels, 0, $\pm V_C$, and $\pm 2V_C$, since the 0 levels of each stage do not sum a different level. Continuing with this reasoning, $(n-1)/2$ stages are required to obtain an n-level leg voltage v_{io}. Assuming that all stages have the same value of DC voltage source V_C, then v_{io} is the sum of all the contributions of the $(n-1)/2$ stages, resulting in

$$v_{io} = V_C \cdot \sum_{j=1}^{\frac{(n-1)}{2}} \left(s_{(2j-1)} - s_{2j} \right) \tag{2.8}$$

In this topology the leg voltage is the same as the load voltage since the load is connected between both terminals of the leg. It is also the phase or line voltage of a multiphase topology, depending on a star or triangle connection of the load.

2.4.1 Five-Level CCMC

Figure 2.19a shows the simplest structure of a CCMC. It is formed with two cascaded stages and corresponds to a five-level cascaded cell multilevel topology. The phase voltage (v_{io}) can be calculated with (2.8) and results in

$$v_{io} = v_1 + v_2 = (s_1 - s_2 + s_3 - s_4)V_C \tag{2.9}$$

The same as the five-level topologies with a common DC bus (5L-FCMC and 5L-DCMC), the 5L-CCMC uses four switching functions and has 16 states. The graphic representation, shown in Figure 2.19b, is realized with two concentric cubes, each one identified with the different values of s_4. The external cube corresponds to $s_4 = 0$, while the inner one is related to $s_4 = 1$. This concentric distribution of the cubes is performed to obtain an adjacent state, which derives in only one voltage level change, between the corners of both cubes.

The corners of each cube in Figure 2.19b are identified by its state and the corresponding voltage level, as calculated with (2.9). There exist redundant states for every voltage level except those extremes $+2V_C$ and $-2V_C$. There exist six states for 0 V (0000, 1001, 1100, 0011, 0110, and 1111), four states for V_C (1011, 0010, 1110, and 1000), and four states for $-V_C$ (0111, 0001, 1101 and 0100).

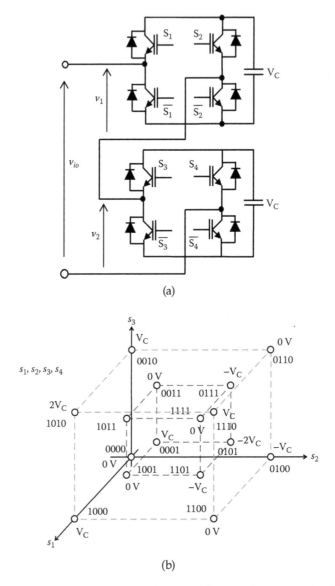

Figure 2.19 Five-level CCMC: (a) topology and (b) state diagram and voltages.

2.5   Summary of Symmetric Topologies

The main characteristics and the component requirements of the *n*-level symmetric topologies, DCMC, FCMC, and CCMC, are summarized in Table 2.1. All of them have a modular structure, so it is easy to increase the number of voltage levels, just adding more stages. All of them have the same number of active switches $2(n-1)$ per leg (DCMC and FCMC) or

Table 2.1 Summary of Symmetric Topologies Characteristics

Voltage Source	Common DC Bus		Isolated
n-Level Topology	FCMC	DCMC	CCMC
Active switches	$2 \cdot (n-1)$ per leg	$2 \cdot (n-1)$ per leg	$2 \cdot (n-1)$ per phase
Equal voltage additional diodes	None	$(n-1) \cdot (n-2)$	None
Equal voltage Capacitors	• $(n-1)$ on DC bus • $(n-2) \cdot (n-1)/2$ per leg	$(n-1)$ on DC bus	$(n-1)/2$ per phase
Modular structure	YES	YES	YES

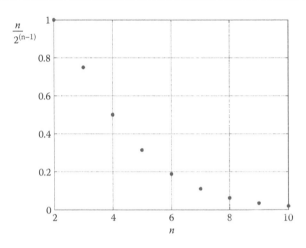

Figure 2.20 Number of voltage levels to topology states ratio versus number of voltage levels.

phase (CCMC). So, there are $(n-1)$ pairs of complementary switches that are controlled by $(n-1)$ switching functions. This determines the existence of $2^{(n-1)}$ possible states in each leg. Since there are more possible states than voltage levels, this means that there are redundant states that generate the same voltage levels. Figure 2.20 shows the ratio between the generated voltage levels with respect to the possible states versus the number of voltage levels. It is clearly seen that this ratio diminishes as n increases.

Table 2.1 also shows the quantity of extra components that are required by each symmetric topology. The DCMC requires $(n-1)(n-2)$ clamping diodes per leg to synthesize n voltage levels. This number can be reduced to $2(n-2)$ per leg if the diodes have different blocking voltage, as shown in Figure 2.13b [6]. The FCMC does not need clamping diodes, but it requires $(n-2) \cdot (n-1)/2$ extra capacitors per leg, plus $(n-1)$ capacitors in the DC link, the same as the DCMC. Each stage can be implemented with a single flying capacitor. In this case the number of extra capacitors is reduced to $(n - 2)$ per leg, but they have to withstand different voltages. The CCMC only requires $(n - 1)/2$ extra capacitors, one in each stage, but it requires isolated DC voltage for each stage.

It is desired to find a topology for which the number of available states is similar to the number of generated voltage levels. The analyzed symmetric topologies are far from meeting this goal, as shown in Figure 2.20. Four-level topologies need the double of possible states, while five-level topologies need 16 states, which is more than three times the number of voltage levels.

2.6   Asymmetric Topologies

The analysis in the previous section indicated that an important characteristic of the symmetric topologies is their modular structure and the easiness to increase the number of voltage levels. But the price for this is an important increment of extra components, passive or active, which is more than proportional to the increment of voltage levels [10,11]. The increment of components also implies a rise in mounting complexity, and thus in the cost of the power converter. So it is important to look for a different alternative when more than four or five voltage levels are required.

2.6.1   Hybrid Asymmetric Topologies

The hybrid multilevel converter (HMC) is a topology with a cascade of different stages connected in series but with different values of DC voltages, as shown in Figure 2.21. Each stage may be built with a different topology, which in turn generates a different number of voltage levels ($n_1, n_2, ..., n_p$). This fact, together with the different values of the DC sources, generates asymmetric topologies in which there are no equal stages [8,12–14]. The term *hybrid* is related to the modulation of each stage. The asymmetry of the DC voltage sources also determines the power handled by each stage. The stages with higher DC voltage manage higher power than those with a lower DC voltage [15]. Then it is possible to implement the power switches of the different stages with different devices. For example, the high-voltage stages may be implemented with integrated gate-commutated thyristor (IGCT) switching at the network frequency, while the low-voltage stages may be implemented with insulated gate bipolar transistor (IGBT) switching at high frequency with pulse width modulation (PWM).

2.6.1.1   CCMC with Different Values of Voltage Sources

Let us consider a cascade of H-bridges connected in series but with different values of DC voltages. Starting with a CCMC with a given number of stages,

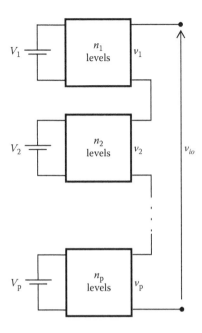

Figure 2.21 Asymmetric topology or HMC.

it is possible to increase the number of voltage levels by adopting different DC voltage sources, which follow certain relationships among them. For example, a two-stage CCMC in which the DC voltage source of one stage doubles the voltage source of the other [13] generates a seven-level HMC (7L-HMC) instead of the 5L-CCMC obtained with equal voltage sources. In a two-stage CCMC with equal DC voltage sources there exist some states in which the voltage generated by one stage cancels the voltage generated by the other. It is not possible to find this situation when different DC voltage sources are used. Moreover, there are new intermediate voltages generated on the phase voltage v_{iN}. The phase voltage of a seven-level HMC can be calculated with (2.8) considering that each stage is fed with a different voltage,

$$v_{io} = v_1 + v_2 = (s_1 - s_2) \cdot V_C + (s_3 - s_4)2 \cdot V_C \qquad (2.10)$$

The 7L-HMC still has four switching functions and 16 possible states, but the number of associated voltage levels is modified. The new voltages are shown in the graphic representation of Figure 2.22a. Comparing the state representation of Figure 2.22a with respect to that of the 5L-CCMC presented in Figure 2.19b, it is possible to find two new voltage levels, $\pm 3V_{DC}$. With the introduction of this asymmetry in the DC voltages there are no longer adjacent states between the corners of both cubes. This means that one level voltage change requires switching more than one switch at a

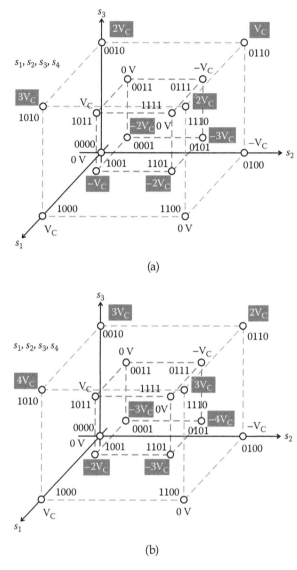

(a)

(b)

Figure 2.22 State diagram of the HMC with different DC buses: (a) $2^{(j-1)}.V_C$ and (b) $3^{(j-1)}.V_C$.

time. This is done with a hybrid modulation. In a p-stage CCMC in which the DC voltages have a progression such that the voltage of stage j equals $2^{(j-1)}.V_C$ (with $j = 1, 2, ..., p$), the number of voltage levels will be $n = 2^{(p+1)} - 1$.

Other commonly used HMCs is that in which the voltage of stage j equals $3^{(j-1)}.V_C$ (with $j = 1, 2, ..., p$), and the resulting number of voltage

Table 2.2 p-Stage CCMC with Different Progressions of DC Voltages and the Resultant AC Voltage Levels

Multilevel Topology	DC Voltage Source of the j Stage	Number of Voltage Levels v_{io}	Level Index
CCMC	$V_j = V_C$	$2p + 1$	$(2p+1)4^{-p}$
Double-progression HMC	$V_j = 2^{(j-1)}V_C$	$2^{(p+1)} - 1$	$2^{(1-p)} - 4^{-p}$
Quasi-triple-progression HMC	$V_j = 2.3^{(j-1)}V_C$	$2.3^{(p-1)} + 1$	$\dfrac{2}{3}\left(\dfrac{3}{4}\right)^p + 4^{-p}$
Triple-progression HMC	$V_j = 3^{(j-1)}V_C$	3^p	$\left(\dfrac{3}{4}\right)^p$

levels is $n = 3^p$. In particular, two stages will have nine voltage levels, as shown in Figure 2.22b. When comparing with Figure 2.19b, there are four new voltage levels, $\pm 3V_{DC}$ and $\pm 4V_{DC}$.

There exist other progressions of DC voltages that have been dealt with in the literature [8,15]. The most popular of them are summarized in Table 2.2, where the CCMC is compared with the HMC with double, quasi-triple, and triple progressions. The table shows the DC voltage progression for a cascade of p stages, the number of voltage levels on v_{io}, and the level index that equals the ratio of the number of voltage levels with respect to the number of possible states.

All the HMCs are CCMC with different values of DC voltage. So all of them are controlled with $2p$ switching functions, and the number of possible states equals 2^{2p} while the phase voltage (v_{io}) is calculated with (2.8) and results in

$$v_{io} = \sum_{j=1}^{p} \left(s_{(2j-1)} - s_{2j} \right) \cdot V_j \tag{2.11}$$

The right column of Table 2.2 illustrates the level index that allows us to evaluate the improvement in voltage levels of one progression or the other for the same number of stages. It is clearly seen that the triple progression is better than all the others. A progression higher than triple does not have practical meaning, since it is not capable of generating symmetrical intermediate states.

2.6.2   Combining Different Topologies

Another way to obtain more voltage levels with fewer stages is combining different multilevel topologies. The combination may contain stages of the same kind, such as a three-level NPC converter or FCMC [16], or different kinds, such as the arrangement of H-bridges with a three-level NPC converter or FCMC [17]. The main goal is to increase the number of voltage levels while decreasing the number of devices, active switches, diodes, and capacitors. There are some new topologies that can behave like the HMC also working with a common DC bus, instead of isolated DC voltage sources.

Figure 2.23a shows a topology with a cascade of different topologies fed with a common DC link [18]. There are two cascaded stages, one with two flying capacitor cells (gray zone) fed by the DC link V_{DC} and the other built with an NPC converter topology. When the switch S_2 is closed, the devices S_1, $\bar{S}_1$, and C_{21} form a three-level FC cell. Similarly, when $\bar{S}_1$ is closed, S_2, $\bar{S}_2$, and C_{22} form another three-level FC cell. If the capacitances of C_{11} and C_{12} are equal, the voltage on them equals $V_{DC}/2$. Then, the voltages on the flying capacitors C_{21} and C_{22} will be $V_{DC}/4$.

Each pair of complementary switches is driven by a different switching function. The same as with the other topologies, it is possible to calculate the leg voltage through the switching functions. The output voltage is the sum of the voltage across $\bar{S}_2(v_{\bar{S}_2})$ plus the voltage generated by the NPC converter stage. Assuming that S_2 is closed ($s_2 = 1$) for a long time, when S_1 is closed the voltage across $\bar{S}_2$ equals the DC link minus the voltage on both flying capacitors, that is, $V_{DC}/2$. When $\bar{S}_1$ is closed the voltage across $\bar{S}_2$ equals half the DC link minus the voltage over the lower flying capacitor, that is, $V_{DC}/4$. Then, it can be expressed as

$$v_{\bar{S}_2} = \left(1 + s_1\right)\frac{V_{DC}}{4} \text{ with } s_2 = 1 \tag{2.12}$$

On the other hand, if $\bar{S}_1$ is closed ($s_1 = 0$) for a long time, the switching of S_2 and $\bar{S}_2$ generates the following voltage across $\bar{S}_2$:

$$v_{\bar{S}_2} = s_2 \frac{V_{DC}}{4} \text{ with } s_1 = 0 \tag{2.13}$$

The states that combine $s_1 = 1$ and $s_2 = 0$ are forbidden, since they close a loop, including all the capacitors, and their original voltages cannot be sustained. Combining the previous results, $v_{\bar{S}_2}$ can be calculated with

$$v_{\bar{S}_2} = \left(s_1 + s_2\right)\frac{V_{DC}}{4} \text{ with } s_1 \neq 1 \text{ or } s_2 \neq 0 \tag{2.14}$$

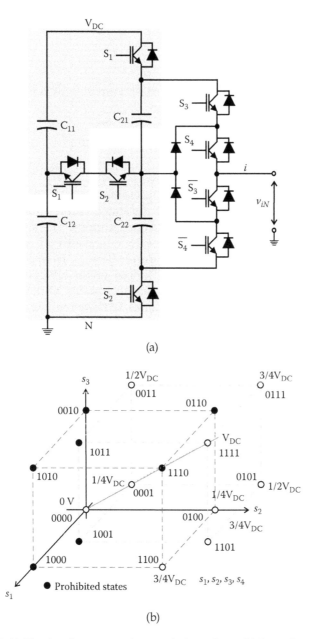

Figure 2.23 (a) Five-level asymmetric cascaded topology. (b) State diagram.

The last stage is an NPC converter with two switching functions, s_3 and s_4, driving the pairs S_3-$\bar{S}_3$ and S_4-$\bar{S}_4$, respectively. Then, the voltage may be expressed as

$$v_{NPC} = \left(s_3 + s_4\right)\frac{V_{DC}}{4} \tag{2.15}$$

Finally, the leg voltage is calculated as the sum of both voltages:

$$v_{io} = \left(s_1 + s_2 + s_3 + s_4\right)\frac{V_{DC}}{4} \tag{2.16}$$

Then the leg voltage has five levels: 0, ¼V_{DC}, ½V_{DC}, ¾V_{DC}, and V_{DC}. Figure 2.23b illustrates the graphic representation of the 16 possible states and the corresponding leg voltages. As stated before, some of the possible states are forbidden for operating conditions. The switches S_1 and $\bar{S}_2$ cannot be closed at the same time independently of the state of the other switches. Assuming a state representation in the order s_1, s_2, s_3, and s_4, all states 10XX should be avoided. Moreover, the NPC converter topology has its own forbidden state (10), and then states XX10 should also be avoided. There are seven forbidden states, so there remain nine allowable states to fix five voltage levels. This means that there are four redundant states.

Even though the ratio between the voltage levels and the total number of states remains the same as that for classical topologies, the number of devices (diodes and capacitors) is reduced when compared to the DCMC or the FCMC. When compared to the DCMC (Figure 2.12a), most of the clamping diodes have been eliminated. When compared to the FCMC (Figure 2.17a), the number of flying capacitors has been reduced. This results in a more compact structure with a reduced mounting complexity.

2.6.3 Cascade Asymmetric Multilevel Converter

Another combination of the cascade connection of two different topologies is presented in Figure 2.24. Here, there is a first stage built with two stacked basic cells and a second stage built with a three-level FCMC [1,3]. This topology is named cascade asymmetric multilevel converter (CAMC). It has two topologically different stages that are fed with different DC voltages, preserving a common DC link. The high-voltage (HV) stage is a two-cell stack similar to the generalized topology; it is fed by the DC link voltage. The low-voltage (LV) stage, connected between nodes w-z, is fed with half the DC link voltage and has the structure of a three-level flying capacitor cell. The capacitors C_1 and C_2 have the same capacitance

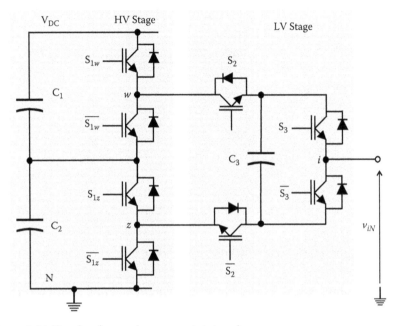

Figure 2.24 Five-level cascade asymmetric topology.

and divide the DC link voltage in two equal shares, $V_{DC}/2$. The switches S_{1w} and S_{1z} use the same switching function (s_1) and the voltage between nodes w-z is fixed to $V_{DC}/2$. So, the LV stage is fed with half the voltage of the HV stage.

The CAMC uses only three switching functions, s_1, s_2, and s_3, to control switches $S_{1(w,z)}$, S_2, and S_3, respectively. Then, there are eight different states to generate five voltage levels. The leg voltage (v_{iN}) is calculated as

$$v_{iN} = \frac{V_{DC}}{2}s_1 + \frac{V_{DC}}{4}(s_2 + s_3) \qquad (2.17)$$

The first term corresponds to the HV stage contribution (0 and $V_{DC}/2$), while the second term corresponds to the LV stage contribution (0, $V_{DC}/4$, and $V_{DC}/2$). Figure 2.25 illustrates the eight available states together with the corresponding leg voltage. All possible states are located at the corners of one cube defined by s_1, s_2, s_3. The switching functions of the LV stage, s_2 and s_3, define the corners of one face of the cube, while s_1 shifts the trajectory between opposite faces of the cube. The back face of the cube is defined by $s_1 = 0$ and the front face by $s_1 = 1$.

Three different levels are encountered when $s_1 = 0$: 0 V for state 000, $V_{DC}/4$ for states 010 ($=V_{C2} - V_{C3}$) and 001 ($= V_{C3}$), and $V_{DC}/2$ for state 011. The other three levels are encountered with $s_1 = 1$: $V_{DC}/2$ for state 100, $3V_{DC}/4$

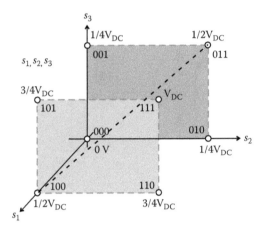

Figure 2.25 CAMC state representation and voltage levels.

($= V_{DC} - V_{C3}$) for state 110, $3V_{DC}/4$ ($=V_{C2} + V_{C3}$) for state 101, and finally V_{DC} for state 111. So, there are two redundant states for each of the intermediate-voltage levels $V_{DC}/4$: $3V_{DC}/4$ and $V_{DC}/2$.

The sequence of the switching states is defined by the modulation strategy, which is employed to move between adjacent states. The neighbor corners of the back and front faces do not represent adjacent states (the voltage difference is higher than one level). Then, the shift from one face to the other is done with the redundant states 100 and 011, for which the leg voltage remains equal to $V_{DC}/2$. This sequence, shown in Figure 2.25 with a bold traced line, requires changing the state of all the switches at the same time. While remaining in $s_1 = 1$ or $s_1 = 0$, the sequence is performed by the LV stage using phase-shifted carrier pulse width modulation (PSPWM). The HV stage switches at the line frequency, so s_1 allows shifting between the front and back faces of the cube at low frequency.

2.7 Summary

A systematic analysis of the classical multilevel topologies analysis together with the introduction of a new method with a graphic representation of the switching states and the voltage levels was presented in this chapter. They allow visualizing and understanding the behavior of multilevel converters. Moreover, they are a powerful tool to develop new topologies with more voltage levels and reduced complexity.

The multilevel topologies are divided in two main groups: (1) those with a common DC source, like the DCMC and the FCMC, and (2) those with multiple and isolated DC sources, like the CCMC. Moreover, they are classified between symmetric, asymmetric, and hybrid topologies. Special attention was given to the capacity of each topology to synthesize different voltages levels with respect to the switching states generated by the different switching functions that control the different stages of each topology. The symmetric topologies have a modular structure where it is easy to add more voltage levels simply by adding more stages of the same structure. But, they present the disadvantage of requiring more power devices, more volume, and higher complexity.

It is possible to diminish the complexity and the device count by generating asymmetric topologies. The first example was the CCMC using different values of DC voltage sources. These are the HMC that utilize specific DC voltage progressions to maximize the number of voltage levels. Another alternative is to use different topologies in the different stages of one multilevel converter. A good example of this is the five-level CAMC, which has only eight possible states to generate five voltage levels. It also has a common DC link that is very attractive for use in back-to-back connection.

References

1. F.Z. Peng. A Generalized Multilevel Inverter Topology with Self Voltage Balancing. *IEEE Transactions on Industry Applications*, 37(2), 611–618, 2001.
2. B.P. McGrath. Topologically Independent Modulation of Multilevel Inverters. PhD thesis, Department of Electrical and Computer Systems Engineering, Monash University, Melbourne, Australia, January 2002.
3. P. Barbosa, P. Steimer, J. Steinke, L. Meysenc, M. Winkelnkemper, N. Celanovic. Active Neutral-Point-Clamped Multilevel Converters. In *IEEE Power Electronics Specialists Conference (PESC'05)*, Recife, Brazil, June 12–16, 2005, pp. 2296–2301.
4. T. Bruckner, S. Bernet. The Active NPC Converter for Medium-Voltage Applications. In *IEEE Industry Applications Annual Meeting (IAS'05)*, Hong Kong, China, October 2–6, 2005, vol. 1, pp. 84–91.
5. A. Nabae, I. Takahashi, H. Akagi. A New Neutral-Point-Clamped PWM Inverter. *IEEE Transactions on Industry Applications*, 17(5), 518–523, 1981.
6. D. Soto, T.C. Green. A Comparison of High-Power Converter Topologies for the Implementation of FACTS Controllers. *IEEE Transactions on Industrial Electronics*, 49(5), 1072–1080, 2002.
7. T.A. Meynard, H. Foch. Multi-Level Conversion: High Voltage Choppers and Voltage-Source Inverters. In *IEEE Power Electronics Specialists Conference (PESC'92)*, Toledo, Spain, June 29–July 3, 1992, vol. 1, pp. 397–403.
8. Y.S. Lai, F.S. Shyu. Topology for Hybrid Multilevel Inverter. *IEE Proceedings on Electric Power Applications*, 149(6), 449–458, 2002.
9. L.M. Tolbert, F.Z. Peng. Multilevel Converters as a Utility Interface for Renewable Energy Systems. In *IEEE Power Engineering Society Summer Meeting (PES'00)*, Seattle, WA, July 16–20, 2000, vol. 2, pp. 1271–1274.
10. K. Fujii, U. Schwarzer, R.W. De Doncker. Comparison of Hard-Switched Multi-Level Inverter Topologies for STATCOM by Loss-Implemented Simulation and Cost Estimation. In *IEEE Power Electronics Specialists Conference (PESC'05)*, Recife, Brazil, June 12–16, 2005, pp. 340–346.

11. W. Hongyang, H. Xiangning. Research on PWM Control of a Cascade Multilevel Converter. In *Power Electronics and Motion Control Conference (IPEMC'00)*, Beijing, China, August 15–18, 2000, vol. 3, pp. 1099–1103.

12. A.A. Sneineh, M. Wang, K. Tian. A Hybrid Capacitor-Clamp Cascade Multilevel Converter. In *IEEE Annual Conference of the Industrial Electronics Society (IECON'06)*, Paris, November 7–10, 2006, pp. 2031–2036.

13. M.D. Manjrekar, T.A. Lipo. A Hybrid Multilevel Inverter Topology for Drive Applications. In *Applied Power Electronics Conference and Exposition (APEC'98)*, Anaheim, CA, February 15–19, 1998, vol. 2, pp. 523–529.

14. M.D. Manjrekar, P.K. Steimer, T.A. Lipo. Hybrid Multilevel Power Conversion System: A Competitive Solution for High-Power Applications. *IEEE Transactions on Industry Applications*, 36(3), 834–841, 2000.

15. J. Dixon, L. Moran. Multilevel Inverter, Based on Multi-Stage Connection of Three-Level Converters Scaled in Power of Three. In *IEEE Annual Conference of the Industrial Electronics Society (IECON'02)*, Seville, Spain, November 5–8, 2002, vol. 2, pp. 886–891.

16. D. Kai, Z. Yumping, L. Lei, W. Zhichao, J. Hongyuan, Z. Xudong. Novel Hybrid Cascade Asymmetric Inverter Based on 5-Level Asymmetric Inverter. In *IEEE Power Electronics Specialists Conference (PESC'05)*, Recife, Brazil, June 12–16, 2005, pp. 2302–2306.

17. C. Rech, J.R. Pinheiro. Hybrid Multilevel Converters: Unified Analysis and Design Considerations. *IEEE Transactions on Industrial Electronics*, 54(2), 1092–1104, 2007.

18. G. Gateau, T.A. Meynard, H. Foch. Stacked Multicell Converter (SMC): Properties and Design. In *Power Electronics Specialists Conference (PESC'01)*, Vancouver, Canada, June 17–21, 2001, pp. 1583–1588.

chapter 3

Diode-Clamped
Multilevel Converter

3.1   Introduction

A generalized method for the analysis of the different multilevel converters has been presented in Chapter 2. One topological derivation from this structure is the diode-clamped multilevel converter (DCMC), whose simplest version, the three-level neutral point clamped (NPC) converter, was introduced in 1981 by Nabae and coworkers [1]. Although the NPC converter has matured to significant industrial acceptance, DCMCs with more than three levels have been relegated to laboratory prototypes mainly due to the problem of DC bus voltage balance. However, its reduced number of capacitors, compared with other multilevel topologies, and its inherent suitability for back-to-back connection still justify the interest to develop new control strategies in order to extend its operation to a higher number of levels.

In this chapter, the analysis is focused on the particular characteristics of the diode-clamped multilevel converter. Its dynamical switching limitations and the voltage balance issue are explained. A suitable control algorithm, which combines the flexibility of multilevel space vector modulation and a DC bus voltage balance control that uses the line voltage redundancy of the three-phase DCMC, is also described.

3.2   Converter Structure and
Functional Description

In Section 2.3.1 the structure of the diode-clamped multilevel converter was derived from the generalized topology. By restructuring the original switching functions it was possible to eliminate redundant capacitors and also active power switches to finally come together with a common DC bus topology. Moreover, a parts-count improvement could also be made toward a more compact and practical implementation by considering different blocking voltages for the clamping diodes. As a counterpart, this configuration has no redundant states for the synthesis of the leg voltages, which is the result of the high number of forbidden states. Besides these properties, an important problem that affects the DCMC is the DC bus voltage balance, which arises in the general case when a multitapped DC voltage source is not available. In this sense, the high number of forbidden states and the nonexistence of redundant states for leg voltage synthesis prevent their use to address voltage balance. In addition, the dynamical switching behavior of the diode-clamped multilevel converter also imposes particular constraints in order to ensure safe operation of the power devices.

3.2.1 Voltage Clamping

The practical implementation of the DCMC derived from the generalized topology of Chapter 2 is depicted in Figure 3.1. It shows one leg of the five-level converter, which is composed of the active switching

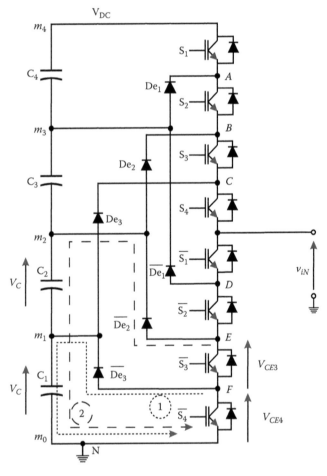

Figure 3.1 Leg of the five-level DCMC, switching logic, and voltage clamping of internal nodes.

devices with their integrated freewheeling diodes, the clamping diodes, and the DC bus with its intermediate nodes. The nodes between the active switches are labeled with the letters A to F in the same figure.

The capability of the DCMC topology to increase the output voltage beyond the maximum blocking voltage of the individual switching devices lies on the voltage-limiting action that the clamping diodes have on the internal nodes of the leg. Figure 3.1 shows this for clamping diodes De_2 and De_3 and the corresponding nodes E and F. Considering that the switches S_3 and S_4 are in the OFF state, it can be seen that the blocking voltage V_{CE4} cannot exceed the value V_C since the anode of De_3 should have lower voltage than its cathode (voltage loop 1 in the figure). In this case the voltage of node F and also V_{CE4} are directly clamped to the capacitor voltage, V_C. In the same way, the voltage of node E is clamped to $2V_C$ due to the diode De_2 (voltage loop 2). However, De_2 does not ensure $V_{CE3} = V_C$ unless V_{CE4} has been previously clamped to V_C. This is true because voltage loop 2 states that $2V_C - V_{CE3} - V_{CE4} = 0$. This dependence of maximum blocking voltage over S_3 is called indirect clamping and applies to nodes B, C, D, and E, and generally to all inner active devices of an n-level DCMC. Only those devices that are located at the top and at the bottom of the leg are directly clamped [2,3]. Therefore, in order to ensure voltage clamping of all switching devices, it is important to previously charge the device output capacitance C_{CE} (in this example C_{CE4}) in each transition. This can be accomplished by following a one-step sequential switching scheme that avoids the simultaneous transition of two adjacent switches. This restriction has been analyzed by Bartolomeüs et al. [4], Adams et al. [5,6], and Hochgraf et al. [7] and is established as a general switching constraint for n-level DCMC converters.

3.2.2 Switching Logic

Figure 2.12b shows the switching functions of a five-level DCMC and it is repeated here for convenience in Figure 3.2. The forbidden and valid states are respectively denoted as solid and hollow circles on each corner of the cubes, and the corresponding leg voltages are specified. As it was mentioned, a great number of forbidden states are observed (11 in total), while the allowed states are reduced to only five, which coincides with the number of converter levels (this is general for the DCMC topology). Table 3.1 summarizes the leg output voltages as a function of the gating signals where it can be seen that no redundancy exists, provided that each output voltage level is synthesized by only one gating pattern. Figure 3.3a–j shows each switching combination of a five-level DCMC for both current flow directions. The gray tone of the gate pins of the active devices denotes their conduction state. The black pin denotes that the device is ON, while the gray pin indicates that the device is OFF. Each

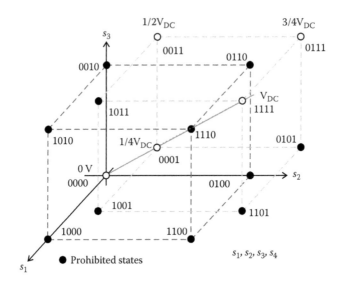

Figure 3.2 State diagram of the five-level DCMC and output voltages.

Table 3.1 Gating Signals and
Corresponding Leg Output Voltages

S_1	S_2	S_3	S_4	v_{iN}
0	0	0	0	0
0	0	0	1	V_C
0	0	1	1	$2\,V_C$
0	1	1	1	$3\,V_C$
1	1	1	1	$4\,V_C$

switching combination acts over the voltage bias of the clamping diodes and the freewheeling diodes, defining a value of leg output voltage (v_{iN}) and a unique path for the output current. The devices through which the current circulates are marked in black, while the remaining devices are in gray. For example in Figure 3.3a, S_1, $\bar{S}_2$, $\bar{S}_3$, and $\bar{S}_4$ are activated. As their complements S_1, S_2, S_3, and S_4 are deactivated, no current can flow through these devices. Also, the freewheeling diodes of S_1, S_2, S_3, and S_4 cannot conduct because of the inverse direction of the load current. On the other hand, although $\bar{S}_1$, $\bar{S}_2$, $\bar{S}_3$, and $\bar{S}_4$ are in the ON state, they cannot conduct the load current. Then, the current finds its path through the freewheeling diodes of $\bar{S}_1$, $\bar{S}_2$, $\bar{S}_3$, and $\bar{S}_4$, forcing the leg voltage to zero. It can be observed that the clamping diodes De_3, De_2, and De_1 are negative biased, provided that their anodes are connected to node N and the

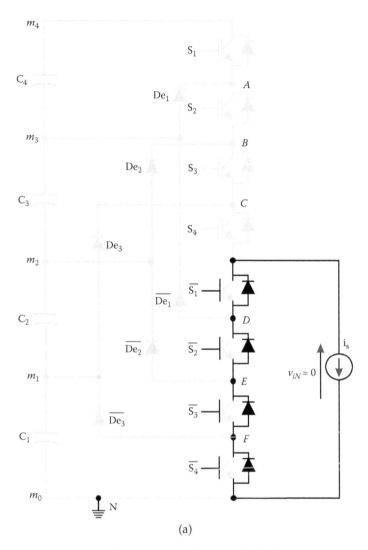

(a)

Figure 3.3 Switching combinations for a five-level DCMC leg with ingoing and outgoing current flows: (a) $v_{iN} = 0$, i_s outgoing; (b) $v_{iN} = 0$, i_s ingoing; (c) $v_{iN} = V_C$, i_s outgoing; (d) $v_{iN} = V_C$, i_s ingoing; (e) $v_{iN} = 2V_C$, i_s outgoing; (f) $v_{iN} = 2V_C$, i_s ingoing; (g) $v_{iN} = 3V_C$, i_s outgoing; (h) $v_{iN} = 3V_C$, i_s ingoing; (i) $v_{iN} = 4V_C$, i_s outgoing; and (j) $v_{iN} = 4V_C$, i_s ingoing.

cathodes are connected to the positive voltages of m_1, m_2, and m_3 of the taps in the DC bus. On the other hand, De_3, De_2, and De_1 clamp the internal nodes C, B, and A to the respective nodes m_1, m_2, and m_3. The same gating scheme is depicted in Figure 3.3b, but the load current is now injected to the leg. In this case the current finds its path through the transistors S_1, S_2,

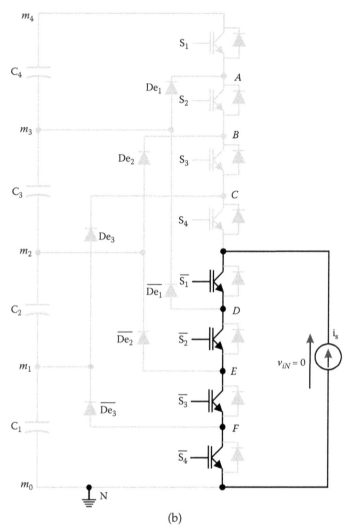

(b)

Figure 3.3 (Continued)

$\bar{S}_3$, and $\bar{S}_4$ because their freewheeling diodes cannot conduct an ingoing current, while the clamping diodes $\overline{De}_3$, $\overline{De}_2$, and $\overline{De}_1$ stay negative biased as in the previous case. The switching combination that synthesizes $v_{iN} = V_C$ is presented in Figure 3.3c,d for both current directions. When the load current is in the outgoing direction, neither $\bar{S}_1$, $\bar{S}_2$, nor $\bar{S}_3$ can conduct, because although they are activated, it is impossible for them to carry current in the inverse direction. Moreover, their associated freewheeling diodes are inversely biased so the load current finds its path through $\bar{S}_4$ and the clamping diode De_3. This connects the load to C_1, establishing that

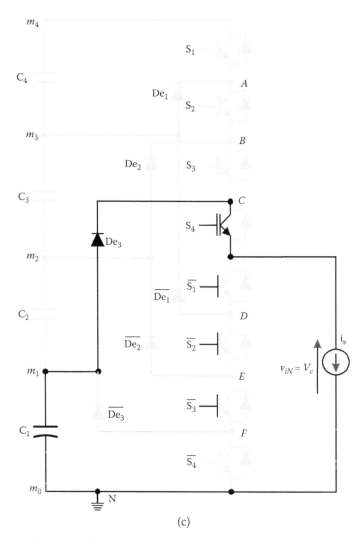

(c)

Figure 3.3 (Continued)

$v_{iN} = V_C$. Similarly, when the current is in the ingoing direction, it can only circulate through S_1, S_2, and $\overline{S_3}$, and also through De_3, defining a path to C_1 and again forcing the output voltage $v_{iN} = V_C$.

The remaining states (Figure 3.3e–h) can be analyzed in a similar manner so that a unique path for the output current to the intermediate taps of the DC bus can always be found. Also, in every state, the voltage-limiting action of the clamping diodes can be observed over the internal nodes of the leg (A to F) and the direct and indirect clamping of the power switches can be verified.

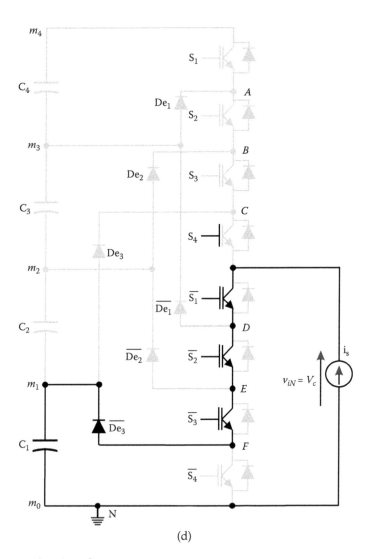

(d)

Figure 3.3 (Continued)

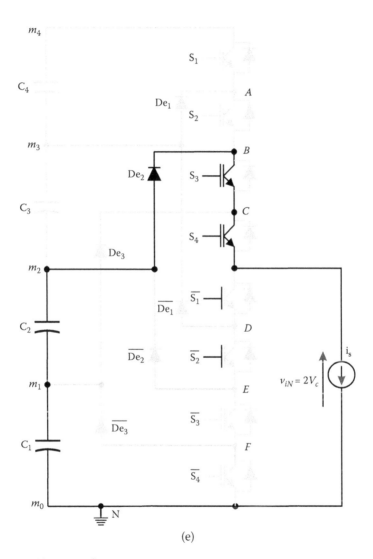

(e)

Figure 3.3 (Continued)

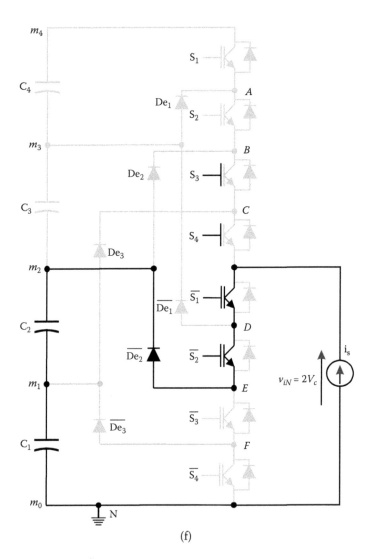

(f)

Figure 3.3 (Continued)

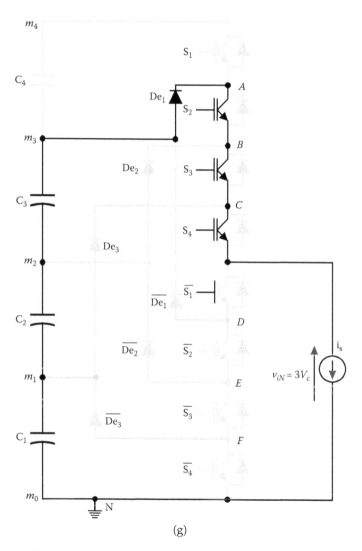

(g)

Figure 3.3 (Continued)

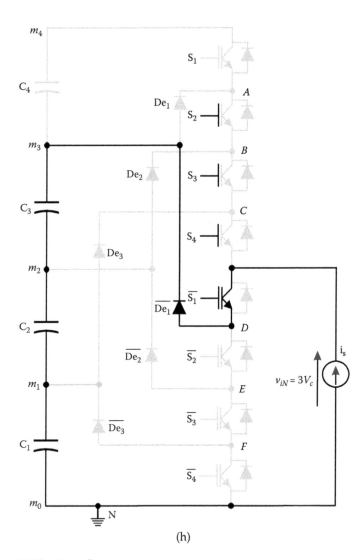

(h)

Figure 3.3 (Continued)

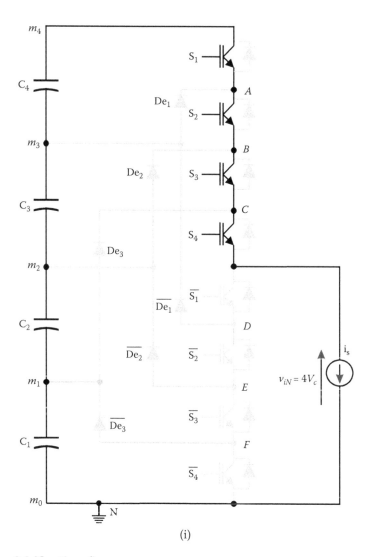

(i)

Figure 3.3 (Continued)

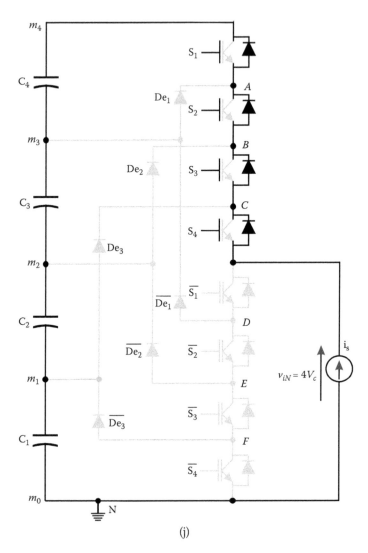

(j)

Figure 3.3 (Continued)

The switching logic of Table 3.1 allows us to synthesize a functional model of the leg that consists of a single-pole multiple-throw switch whose inputs are the intermediate taps of the DC bus, and the cursor defines the leg output voltage. This simplified model for the five-level converter is depicted in Figure 3.4a, jointly with a typical stepped voltage waveform in Figure 3.4b. It is clear that the load current circulates through the cursor of the switch and is injected or drained from the capacitor nodes of the DC bus. The voltage balance problem can be visualized by considering the flow of the output current into the DC bus using the switching pattern

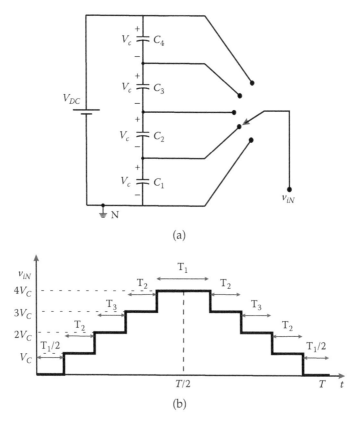

(a)

(b)

Figure 3.4 (a) Simplified model of a five-level DCMC leg. (b) Leg voltage with respect to the negative of the DC bus.

of Figure 3.4b. The switching times (T_1, T_2, T_3) set up the duration of the cursor in each node of the DC bus, thus defining the shape of the voltage waveform. In Figure 3.5, the line current i_i is supplied from node m_1 and produces different voltage deviations, depending on the position of each capacitor on the DC bus. Particularly in this case, the current is drained from m_1 and C_1 is discharged (with respect to the reference voltage V_C), while C_2, C_3, and C_4 tend to overcharge.

An expression of the currents drained from the taps of the capacitor divider can be defined as a function of the phase current and the switch position. The current flowing to each tap of the DC bus is expressed in (3.1), and its average value within a fundamental cycle determines the voltage deviation on each capacitor.

$$i_x = \begin{cases} i_i & \text{if } v_{iN} = V_x \\ 0 & \text{otherwise} \end{cases} \qquad x = 0,1,...,4 \qquad (3.1)$$

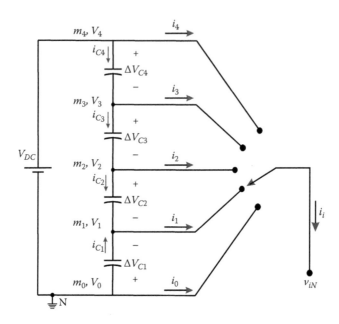

Figure 3.5 Line current i_i causing voltage deviations on DC bus capacitors.

A qualitative assessment about capacitor voltage behavior can be made by looking at Figure 3.6. Figure 3.6a considers the stepped voltage waveform of Figure 3.4b and a sinusoidal current with a phase shift of 90° leading. It is clearly seen that within the time interval $0 < t < T/2$ a positive current is injected into node m_1 ($v_{iN} = V_1$), while a negative pulse occurs when v_{iN} assumes the value V_1 during $T/2 < t < T$. The net charge injected into node m_1 is zero along the period T, and so is the capacitor voltage variation. The same conclusion yields for the remaining nodes, m_2 and m_3. Moreover, this is also true when 90° lagging current is considered. This allows asserting that, in ideal conditions, the processing of pure reactive current has no net effect on capacitors' voltage balance. Figure 3.6b shows the case where active power is transferred from the voltage source V_{DC} to the AC side due to an output current with positive power factor (PF). As can be seen by summing up the contributions during $0 < t < T/2$ and $T/2 < t < T$, a negative average current is injected to m_1 ($v_{iN} = V_1$). This implies that unlike the previous case, there is no charge balance condition over node m_1, and consequently, a net voltage variation of V_1 is expected. Moreover, by virtue of the half-wave symmetry of the synthesized voltage v_{iN}, two additional observations can be pointed out from this figure: (1) for any relative phase shift of the output current the average current entering node m_2 is zero ($\langle i_2 \rangle = 0$), and (2) the average current that is drained from node m_3 is positive and equal to the average current that is injected into node m_1, that is, $\langle i_3 \rangle > 0$ and $\langle i_1 \rangle = -\langle i_3 \rangle$.

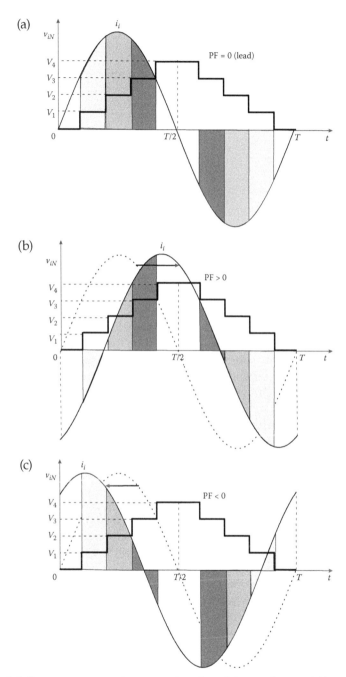

Figure 3.6 Current entering the internal nodes of the DC bus at different phase shifts between output current and leg voltage: (a) PF = 0, (b) PF > 0 ($P_{DC} \rightarrow P_{AC}$), and (c) PF < 0 ($P_{AC} \rightarrow P_{DC}$).

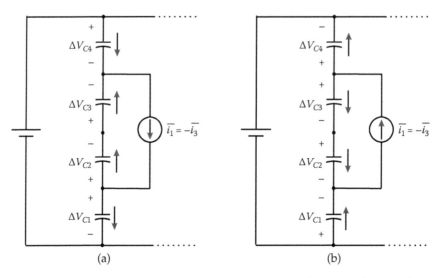

Figure 3.7 Voltage variation on DC bus capacitors as a function of power flow direction: (a) PF > 0 and (b) PF < 0.

On the other hand, Figure 3.6c shows the case where the power factor is negative; that is, power flows from the AC side to the voltage source V_{DC}. This figure shows that the average current on the midpoint tap is zero ($\langle i_2 \rangle = 0$), and that the average current along T $\langle i_1 \rangle$ is positive, while $\langle i_3 \rangle$ is negative. Similar to the previous case, this means that $\langle i_1 \rangle > 0$ and $\langle i_3 \rangle = -\langle i_1 \rangle$. Based on this analysis, an equivalent circuit that describes the behavior of the voltages across the capacitors can be stated when active power flows from DC to the AC side, and vice versa. Figure 3.7a shows the average current flow when power flows from V_{DC} to the AC side (PF > 0, inverter operation). The direction of the load current produces a persistent voltage increment on C_1 and C_4 and a voltage decrement on C_2 and C_3. On the other hand, when the converter is in rectifier mode (PF < 0) (Figure 3.7b), C_1 and C_4 suffer a constant voltage decrement and the opposite occurs with C_2 and C_3 (a generalized analysis regarding the voltage distribution of charge/discharge of the DC bus capacitors can be found in Marchesoni et al. [8]).

The previous discussion yields to the conclusion that a single DCMC leg is not capable to synthesize a desired sequence of output voltages and to maintain the balance on the DC bus simultaneously. The exception to this assertion arises when the power factor is strictly zero. On the other hand, when a positive or negative power factor is considered, the capacitors are charged/discharged symmetrically with respect to the midpoint tap of the DC bus. Particularly when the converter operates as inverter ($P_{DC} \rightarrow P_{AC}$), the inner capacitors are discharged and the outer capacitors are overcharged, while the opposite occurs when the rectifier operation is

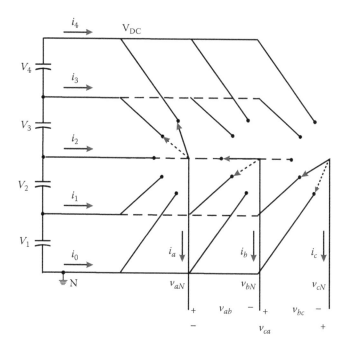

Figure 3.8 Line voltage redundancy and current flow on DC bus nodes.

established $(P_{AC} \rightarrow P_{DC})$; that is, the inner capacitors are overcharged and the outer capacitors are discharged.

Summarizing, there is no way to influence over the voltage balance of the capacitors by individually considering the converter legs, since for a desired output voltage there is a unique destination for the load current, and therefore a unique set of capacitor voltage deviations. To overcome this, several proposals have been presented based on additional converters that basically inject current between the internal nodes of the bus to counteract the natural unbalancing process [9,10]. However, a different prospect arises when considering the line voltage redundant states of the three-phase DCMC driving three-phase three-wire loads. This can be understood by looking at Figure 3.8, where a functional model of the three-phase DCMC is depicted. In this case, the three legs simultaneously inject and drain current to and from the intermediate nodes of the DC bus. This means that the voltage variations on all capacitors are the result of the contribution of the three line currents, the cursor positions, and the duration of each switching combination. As the load is considered without neutral connection, it can slide along the taps of the DC bus while keeping invariant the set of line voltages. That is, it is possible to synthesize a given set of line voltages (v_{ab}, v_{bc}, v_{ca}) with different sets of leg voltages (v_{aN}, v_{bN}, v_{cN}), and although they have the same effect from the point of view of the load, they have different impacts on the net charge injected into the nodes of the DC bus. For example, the two possible switching states, indicated with solid and dotted lines in Figure 3.8, are completely equivalent from the point of view of a three-wire load. However, for the solid combination, $i_4 = i_a$, $i_2 = i_b$, $i_1 = i_c$, and $i_3 = i_0 = 0$, while for the dotted combination, $i_3 = i_a$, $i_1 = i_b$, $i_0 = i_c$, and $i_4 = i_2 = 0$. This fact suggests that for a given set of line voltages, there may be some combination of leg voltages that better contributes to the voltage balance. Of course, the availability of redundant configurations increases as the modulation index decreases, reaching no redundancy when any line voltage assumes the value V_{DC} or $-V_{DC}$ and maximum redundancy when zero line voltage has to be applied to the load, since there are five combinations of leg voltages (v_{aN}, v_{bN}, v_{cN}) that synthesizes zero line voltage, as the three legs can be connected together to V_0, V_1, ..., V_4.

At this point we can summarize the main aspects that should be taken into account for control of a three-phase DCMC driving three-phase three-wire loads:

1. A modulation strategy is necessary for the determination of the line voltages to be synthesized by the converter and to be applied to the load.
2. A switching strategy is necessary to control the voltage balance of the DC bus respecting the previously determined line voltages combinations.
3. The entire control algorithm should have enough flexibility to incorporate the operational switching constraints explained in Section 3.1.

3.3   *Modulation of Multilevel Converters*

The modulation of switching converters has the objective of reproducing a continuous reference signal from a digital pulse train whose average value coincides with the reference. For multilevel converters, several methods have been developed that are, in general, extensions from already known techniques. In a general way, these methods can be classified depending on the switching frequency of the power devices as line frequency switching and high-frequency switching modulators, as shown in Figure 3.9. These strategies have different performances regarding losses, harmonic distortion, implementation complexity, and flexibility for variable speed operation.

Among the high-frequency switching strategies, the subharmonic modulation with shifted carriers was the first extension to multilevel converters due to simplicity and ease of implementation. Two variants of this method are the level-shifted and the phase-shifted carrier modulations, which have found application in different multilevel topologies. The level-shifted carrier modulation was mainly applied to the DCMC converter

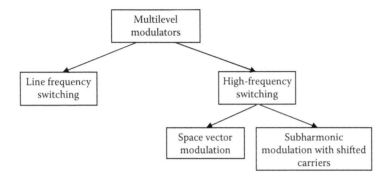

Figure 3.9 Multilevel modulators.

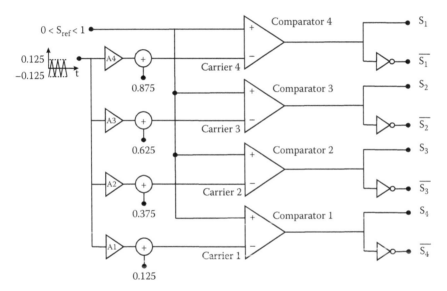

Figure 3.10 Implementation of a level-shifted carrier PWM scheme for a five-level DCMC with comparators.

and consists of the comparison of the modulating reference signal with a set of $(n-1)$ triangular carriers associated with the n levels of the converter. The gating signals S_1, S_2, S_3, and S_4 are the result of the comparison between the modulating signal S_{ref} and the corresponding carriers, which are shifted in level to span the complete range of S_{ref} as shown in Figure 3.10. Different phase shifts can be introduced to the triangular carriers defining three basic switching patterns. This is carried out by setting A_1, A_2, A_3, and A_4. One of them is called phase disposition pulse width modulation (PD-PWM), and it is achieved by setting $A_1 = A_2 = A_3 = A_4 = 1$. The reference signal and its comparison with the carriers are shown in Figure 3.11a, jointly with the resulting leg voltage. In addition, two schemes have been presented: phase opposition disposition (POD-PWM) and alternative phase opposition disposition (APOD-PWM). The relative phase displacement for POD-PWM is achieved when $A_1 = A_2 = 1$ and $A_3 = A_4 = -1$, while APOD-PWM is achieved with $A_1 = A_3 = 1$ and $A_2 = A_4 = -1$. Figure 3.11b,c shows POD and APOD results, respectively.

This modulation technique has been extensively studied, and it is demonstrated through spectral analysis that the variant PD-PWM has the best harmonic performance because it places harmonic energy in a common mode first-carrier component, which is canceled in the line-to-line voltages. On the other hand, although this method is extremely efficient to synthesize multilevel voltages with low hardware complexity, its main disadvantage lies in the fact that it synthesizes the line voltages by directly defining the leg voltages. This fact rigidifies the modulation scheme in the sense that it does not allow the decoupling of the line voltage synthesis from the leg voltages, thus preventing the use of redundant states of the three-phase converter for voltage balancing control of the DC bus. In this sense, a suitable method to control the DCMC driving a three-wire load should allow us to calculate the set of three line voltages (v_{ab}, v_{bc}, v_{ca}) of the load and then select the most adequate realization (v_{aN}, v_{bN}, v_{cN}) in order to keep the voltage balance on the DC bus capacitors. In the next section a multilevel space vector modulator is explained. This modulator synthesizes the correct set of line voltages at the load terminals, but leaving the determination of v_{aN}, v_{bN}, and v_{cN} to a subsequent process that optimizes the DC bus voltage balance.

3.3.1 Multilevel Space Vector Modulation

Space vector modulation is a well-known modulation scheme that has been widely adopted for the control of two-level VSC due to its high flexibility. It is particularly interesting when a fast dynamic response or variable frequency operation is required, for example, high-performance drives and power conditioning systems. Also, it intrinsically has a high DC bus voltage utilization factor and can also be naturally implemented on digital hardware [11]. Redundant states of the three-phase converter can be explicitly exploited, allowing us to calculate the line voltages independently of leg voltages. Nevertheless, the extension of two-level SVM to multilevel converters is a challenging task, mainly due to the existence of a high number of switching states. A brief introduction to multilevel space vector representation is given in the following paragraphs.

A widely accepted representation of the space vector is defined in terms of the phase voltages v_{ao}, v_{bo}, and v_{co}. Generally, these are transformed to the three space vector components v_{α}, v_{β}, and v_0 through a coordinate transformation $T_{\alpha\beta 0}$, according to (3.2):

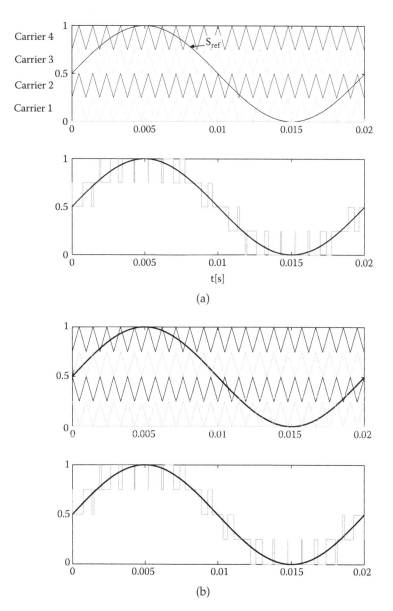

Figure 3.11 Multilevel subharmonic modulation. Triangular carriers and modulating signal for (a) PD-PWM, (b) POD-PWM, and (c) APOD-PWM.

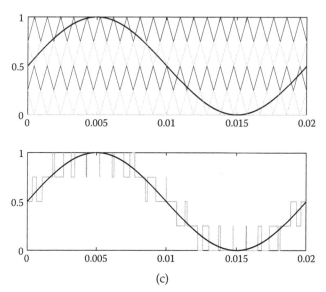

(c)

Figure 3.11 (Continued)

$$
\begin{bmatrix} v_\alpha \\ v_\beta \\ v_0 \end{bmatrix} = T_{\alpha\beta0} \begin{bmatrix} v_{ao} \\ v_{bo} \\ v_{co} \end{bmatrix} \quad \text{where:} \quad T_{\alpha\beta0} = \frac{2}{3} \begin{bmatrix} 1 & -\dfrac{1}{2} & -\dfrac{1}{2} \\ 0 & \dfrac{\sqrt{3}}{2} & -\dfrac{\sqrt{3}}{2} \\ \dfrac{1}{\sqrt{2}} & \dfrac{1}{\sqrt{2}} & \dfrac{1}{\sqrt{2}} \end{bmatrix} \tag{3.2}
$$

This representation allows us to describe the converter switching states with one vector in α-β-0 coordinates. Figure 3.12a shows the simplified model of a standard two-level voltage source converter. In this case, the leg voltage can assume only two values, $m = 1$ and $m = 0$, and therefore six active vectors can be synthesized on the α-β plane, as shown in Figure 3.12b.

Each vector in Figure 3.12b is identified according to m for the three phases ($m_a\, m_b\, m_c$). A particular case appears for vectors $\mathbf{SV}_0$ and $\mathbf{SV}_7$ (zero vectors), where two sets of $m_a, m_b,$ and m_c result in the same representation. $\mathbf{SV}_0$ is synthesized connecting the three phases to 0 V ($m_a = m_b = m_c = 0$), while $\mathbf{SV}_7$ is obtained with $m_a = m_b = m_c = 1$. Since both states are equivalent from the point of view of a three-wire load, the selection of the zero vector ($\mathbf{SV}_0$ or $\mathbf{SV}_7$) is commonly used for the reduction of the switching frequency. Summarizing, the two-level converter has $2^3 = 8$ switching combinations, while the number of different synthesizable vectors equals 7.

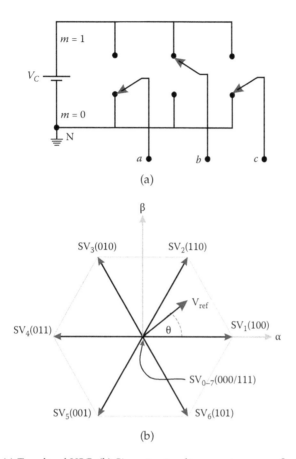

Figure 3.12 (a) Two-level VSC. (b) Six output voltage vectors on α-β plane.

Looking back to Figure 3.12b, the space vector modulation is based on the approximation of the reference vector $\mathbf{V}_{ref}$ through the averaging of the nearest three vectors (NTV) within a sampling period T_S,

$$\mathbf{V}_{ref} = d_1 \, \mathbf{V}_1 + d_2 \, \mathbf{V}_2 + d_3 \, \mathbf{V}_3 \qquad (3.3)$$

where $\mathbf{V}_1$ and $\mathbf{V}_2$ are adjacent vectors that are directly identified from the phase of the reference vector (θ), $\mathbf{V}_3$ is one zero vector, and d_1, d_2, and d_3 are the corresponding duty cycles for each vector. Figure 3.13 shows an averaging period, which is divided into the three activation times corresponding to $\mathbf{V}_1$, $\mathbf{V}_2$, and $\mathbf{V}_3$.

Now considering the simplified model introduced in Figure 3.4a for the three-level DCMC (Figure 3.14), it can be seen that each leg can assume three different values: $m = 0, 1, 2$. The vector map is significantly extended

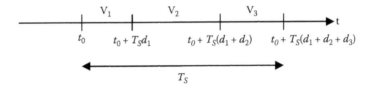

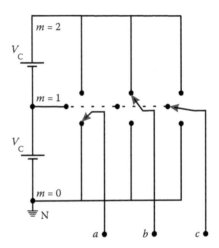

Figure 3.13 Averaging period.

Figure 3.14 Three-level DCMC.

(Figure 3.15). The labels at the tip of each vector indicate the corresponding sets of m. From Figure 3.15, it follows that several vectors can be obtained with more than one combination of leg voltages. More precisely, the vectors that define the internal hexagon can be synthesized by two different combinations, while the zero vector has three possible switching combinations. This redundancy is summarized as 19 different vectors with $3^3 = 27$ possible switching combinations.

Generally speaking, for an n-level DCMC, there exist n^3 switching combinations, while the number of synthesizable vectors is given by the following expression [11]:

$$L = 1 + 6\sum_{i=1}^{n-1} i \tag{3.4}$$

As n increases, the number of hexagonal rings and redundant states also increases. Besides, for $n > 2$, the nearest three vectors cannot be determined only by means of the reference vector angle θ. In this case, there exists a complex relationship between the amplitude and phase

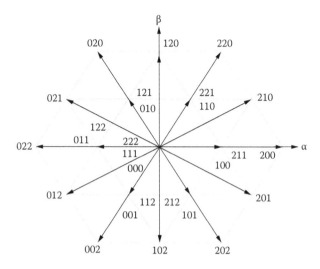

Figure 3.15 Vector map for a three-level DCMC.

of the reference vector to uniquely determine the nearest three vectors. Therefore, the NTV selection is not a simple task, and in some sense, the representation of converter vectors in an orthogonal reference system does not seem to be the best choice to facilitate their identification. A decomposition method of the internal hexagons is developed and applied to the NPC converter reported by Suh et al. [12]. Although the proposal is general for an arbitrary number of levels, the computational effort increases dramatically with the number of voltage levels. However, it has been demonstrated by Joetten and Kehl [13] for three levels and by Celanovic and Boroyevich [14] for n levels that a vector representation in a hexagonal reference system can alleviate the computational burden. Particularly, an adequate coordinate transformation can lead to integer components for all converter output vectors.

3.3.1.1 *Hexagonal Coordinate System*

The identification of the NTVs in a multilevel vector map as part of the synthesis of a given reference vector depends jointly on their angle and modulus. Also, the high number of switching states (especially when n is high) and the inherent switching constraints of the DCMC make this a difficult process, especially when an efficient management of the computational resources is also mandatory. In this sense, it has been demonstrated in Marchesoni and Tenca [14] that a big computational effort can be avoided by using a hexagonal coordinate system for the representation of the space vector, instead of using the most common α-β-0 transformation. This representation is geometrically more consistent than the α-β-0 representation and allows an immediate identification of the nearest three vectors.

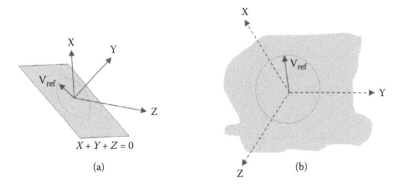

Figure 3.16 (a) 3D coordinate system. (b) Projection of the coordinate axes and the reference vector over the plane $X + Y + Z = 0$.

First, a line voltage vector is defined instead of using phase voltages. This gives direct information of the output state of the converter, with independence of the type of load and its internal connection. A vector **V** is represented in a 3D space (**X**, **Y**, **Z**), and its components are defined as the three line voltages of a three-phase system:

$$\mathbf{V} = v_{ab}\mathbf{X} + v_{bc}\mathbf{Y} + v_{ca}\mathbf{Z} \quad \text{or:} \quad \mathbf{V} = [v_{ab} \ v_{bc} \ v_{ca}]^T \tag{3.5}$$

Given that line voltages (v_{ab}, v_{bc}, v_{ca}) sum zero, the vector (3.5) always exists within the plane $X + Y + Z = 0$. Also, a set of sinusoidal balanced voltages can be represented as a reference vector:

$$\mathbf{V}_{ref} = V_{llpeak} [\cos(\omega t) \quad \cos(\omega t - 2\pi/3) \quad \cos(\omega t + 2\pi/3)]^T \tag{3.6}$$

where V_{llpeak} is the peak line voltage and ω the rotational speed of $\mathbf{V}_{ref}$ within the plane $X + Y + Z = 0$, as shown in Figure 3.16. For any value of ω, the vector defined in (3.6) is contained in a plane, and therefore it could be represented by two components instead of three.

The design of a similarity transformation that changes the coordinates of a given point in the (X, Y, Z) space to the hexagonal coordinate system begins with the definition of a set of three vectors, namely, **u**, **v**, and **z**, whose components in (X, Y, Z) are

$$[\mathbf{u}, \ \mathbf{v}, \ \mathbf{z}] = \begin{bmatrix} V_C & 0 & V_C \\ 0 & V_C & V_C \\ -V_C & -V_C & V_C \end{bmatrix} \tag{3.7}$$

Then, a coordinate transformation is defined that changes the (**X**, **Y**, **Z**) coordinates to another set (**g**, **h**, **y**), where **u** becomes **g** = $[1\,0\,0]^T$, **v** becomes **h** = $[0\,1\,0]^T$, and **z** becomes **y** = $[0\,0\,1]^T$. In this coordinate system, the output vectors of any multilevel converter, with voltage steps equal to V_C, have integer components in **g** and **h** directions and zero in the **y** direction. The transformation matrix P is obtained solving the following equality:

$$[\mathbf{g},\mathbf{h},\mathbf{y}] = \begin{bmatrix} 1 & 0 & 0 \\ 0 & 1 & 0 \\ 0 & 0 & 1 \end{bmatrix} = P[\mathbf{u},\mathbf{v},\mathbf{z}] = P \begin{bmatrix} V_C & 0 & V_C \\ 0 & V_C & V_C \\ -V_C & -V_C & V_C \end{bmatrix}$$

(3.8)

$$\text{with: } P = \begin{bmatrix} V_C & 0 & V_C \\ 0 & V_C & V_C \\ -V_C & -V_C & V_C \end{bmatrix}^{-1} = \frac{1}{3V_C}\begin{bmatrix} 2 & -1 & -1 \\ -1 & 2 & -1 \\ 1 & 1 & 1 \end{bmatrix}$$

Therefore, the coordinate transformation yields to

$$\begin{bmatrix} V_g \\ V_h \\ V_y \end{bmatrix} = \frac{1}{3V_C}\begin{bmatrix} 2 & -1 & -1 \\ -1 & 2 & -1 \\ 1 & 1 & 1 \end{bmatrix}\begin{bmatrix} v_{ab} \\ v_{bc} \\ v_{ca} \end{bmatrix}$$

(3.9)

As an example, a four-level DCMC is considered in Figure 3.17, and the hexagonal representation of the output vector (**g**, **h**, **y**) is calculated applying the P transform:

$$\begin{bmatrix} V_g \\ V_h \\ V_y \end{bmatrix} = \frac{1}{3V_C}\begin{bmatrix} 2 & -1 & -1 \\ -1 & 2 & -1 \\ 1 & 1 & 1 \end{bmatrix}\begin{bmatrix} -2V_C \\ 3V_C \\ -V_C \end{bmatrix} = \begin{bmatrix} -2 \\ 3 \\ 0 \end{bmatrix}$$

Given that the component in the **y** direction is zero for any set of line voltages, P can be simplified:

$$P' = \frac{1}{3V_C}\begin{bmatrix} 2 & -1 & -1 \\ -1 & 2 & -1 \end{bmatrix}$$

(3.10)

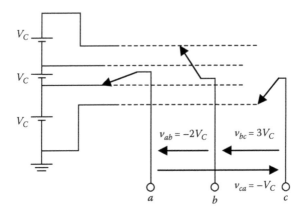

Figure 3.17 Four-level DCMC.

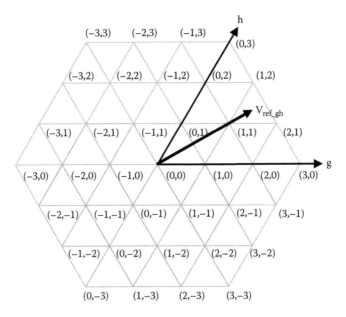

Figure 3.18 Four-level vector map in (g, h) plane.

The coordinate transformation P normalizes all vector components to the step voltage V_C. According to this, the tips of converter vectors in the g-h plane define the corners of triangles depicted in Figure 3.18, which is the map of synthesizable vectors of the four-level converter. By virtue of the normalization to V_C, all vertexes have integer components, as is indicated in bracketed pairs (g, h). On the other hand, a balanced sinusoidal three-phase system generates a rotating vector describing a circular trajectory at the system frequency. The components of this reference vector are calculated using P' as

$$\mathbf{V}_{ref\_gh} = P'\mathbf{V}_{ref} = \begin{bmatrix} V_{ref\_g} & V_{ref\_h} \end{bmatrix}^T \tag{3.11}$$

The modulator should determine the proper set of line voltages in order to synthesize $\mathbf{V}_{ref\_gh}$ by means of (3.3). The algorithm comprises two stages:

1. Nearest three vectors identification
2. Duty cycle calculation

3.3.1.2 Nearest Three Vectors Identification

Figure 3.19 shows a detail of the first sextant of Figure 3.18. The reference vector and its components are depicted, where its tip falls inside the parallelogram defined by the vertexes (2, 2), (1, 2), (1, 1), and (2, 1).

A straightforward determination of the four nearest vectors is accomplished by rounding the components of $\mathbf{V}_{ref\_gh}$. The vectors that define the parallelogram are assigned with names according to the rounding of their components g and h, depending on if they are up-rounded (*upper*) or down-rounded (*lower*). For example, the up-rounded component of V_{ref_g} ($V_{ref\_g}\uparrow$) of Figure 3.19 is 2, whereas the down-rounded component ($V_{ref\_g}\downarrow$) is 1. Then, the vectors that define the parallelogram can be expressed as rounded $\mathbf{V}_{ref\_gh}$ components, as follows:

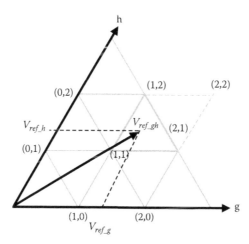

Figure 3.19 Reference vector and its nearest vectors.

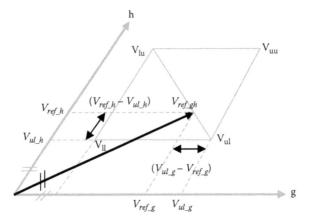

Figure 3.20 Third nearest vector determination.

$$\mathbf{V}_{lu} = [V_{ref\_g} \downarrow \quad V_{ref\_h} \uparrow]^T \qquad \mathbf{V}_{uu} = [V_{ref\_g} \uparrow \quad V_{ref\_h} \uparrow]^T$$

$$\mathbf{V}_{ll} = [V_{ref\_g} \downarrow \quad V_{ref\_h} \downarrow]^T \qquad \mathbf{V}_{ul} = [V_{ref\_g} \uparrow \quad V_{ref\_h} \downarrow]^T \qquad (3.12)$$

In particular for the reference vector of Figure 3.19,

$$\mathbf{V}_{lu} = \begin{bmatrix} 1 \\ 2 \end{bmatrix} \quad \mathbf{V}_{uu} = \begin{bmatrix} 2 \\ 2 \end{bmatrix} \quad \mathbf{V}_{ll} = \begin{bmatrix} 1 \\ 1 \end{bmatrix} \quad \mathbf{V}_{ul} = \begin{bmatrix} 2 \\ 1 \end{bmatrix}$$

The definition of the three nearest vectors begins with the direct selection of $\mathbf{V}_{ul}$ and $\mathbf{V}_{lu}$, while the identification of the third vector between $\mathbf{V}_{uu}$ and $\mathbf{V}_{ll}$ results from the comparison of the components of $\mathbf{V}_{ul}$ and $\mathbf{V}_{ref\_gh}$. With reference to Figure 3.20, the sign of the number W in (3.13) can be used to select $\mathbf{V}_{uu}$ or $\mathbf{V}_{ll}$:

$$W = V_{ref\_h} + V_{ref\_g} - V_{ul\_g} - V_{ul\_h} \qquad (3.13)$$

where V_{ul_g}, V_{ul_h}, V_{ref_g}, and V_{ref_h} are the components of vectors $\mathbf{V}_{ul}$ and $\mathbf{V}_{ref}$. $W = 0$ represents a straight line that links the corners $\mathbf{V}_{ul}$ and $\mathbf{V}_{lu}$. In this case, the averaged vectors are only $\mathbf{V}_{lu}$ and $\mathbf{V}_{ul}$. $W < 0$ implies that the tip of the reference vector falls inside the triangle defined by the vertexes $\mathbf{V}_{lu}$, $\mathbf{V}_{ul}$, and $\mathbf{V}_{ll}$, while $W > 0$ indicates that the triangle is defined by $\mathbf{V}_{ul}$, $\mathbf{V}_{lu}$, and $\mathbf{V}_{uu}$:

$$\mathbf{V}_1 = \mathbf{V}_{lu}$$

$$\mathbf{V}_2 = \mathbf{V}_{ul}$$

$$\mathbf{V}_3 = \begin{cases} \mathbf{V}_{ll} & \text{if } W < 0 \\ \mathbf{V}_{uu} & \text{if } W > 0 \\ \text{not used} & \text{if } W = 0 \end{cases} \tag{3.14}$$

3.3.1.3   Duty Cycle Calculation

The duty cycle calculation is made using Equation (3.3), assigning to $\mathbf{V}_1$, $\mathbf{V}_2$, and $\mathbf{V}_3$ the corresponding coordinates of $\mathbf{V}_{ul}$, $\mathbf{V}_{lu}$, and $\mathbf{V}_3$:

$$\mathbf{V}_{ref\_gh} = d_{ul}\,\mathbf{V}_{ul} + d_{lu}\,\mathbf{V}_{lu} + d_3\,\mathbf{V}_3 \tag{3.15}$$

where d_{ul}, d_{lu}, and d_3 are the duty cycles of the corresponding vectors. Thus, when $\mathbf{V}_3 = \mathbf{V}_{uu}$, its duty cycle can be calculated as $d_3 = d_{uu} = 1 - d_{lu} - d_{ul}$, and the expanded version of Equation (3.15) looks like

$$\begin{bmatrix} V_{ref\_g} \\ V_{ref\_h} \end{bmatrix} = d_{ul}\begin{bmatrix} V_{ul\_g} \\ V_{ul\_h} \end{bmatrix} + d_{lu}\begin{bmatrix} V_{lu\_g} \\ V_{lu\_h} \end{bmatrix} + d_{uu}\begin{bmatrix} V_{uu\_g} \\ V_{uu\_h} \end{bmatrix} =$$

$$= d_{ul}\begin{bmatrix} V_{ul\_g} - V_{uu\_g} \\ V_{ul\_h} - V_{uu\_h} \end{bmatrix} + d_{lu}\begin{bmatrix} V_{lu\_g} - V_{uu\_g} \\ V_{lu\_h} - V_{uu\_h} \end{bmatrix} + \begin{bmatrix} V_{uu\_g} \\ V_{uu\_h} \end{bmatrix} \tag{3.16}$$

Also, considering the integer part operator [.], it can be stated that

$$V_{ul\_g} = V_{uu\_g} = [V_{ref\_g}]+1 \ , \ \ V_{lu\_h} = V_{uu\_h} = [V_{ref\_h}]+1, \tag{3.17}$$

$$\text{and: } V_{ul\_h} - V_{lu\_h} = -1 \ , \ \ V_{lu\_g} - V_{uu\_g} = -1$$

Replacing these relations into (3.16),

$$\begin{bmatrix} V_{ref\_g} \\ V_{ref\_h} \end{bmatrix} = d_{ul}\begin{bmatrix} 0 \\ -1 \end{bmatrix} + d_{lu}\begin{bmatrix} -1 \\ 0 \end{bmatrix} + \begin{bmatrix} [V_{ref\_g}]+1 \\ [V_{ref\_h}]+1 \end{bmatrix} \tag{3.18}$$

Solving (3.18) for d_{ul}, d_{lu}, and d_{uu} results in

$$d_{ul} = 1 - \left(V_{ref\_g} - \left[V_{ref\_g} \right] \right)$$

$$d_{lu} = 1 - \left(V_{ref\_h} - \left[V_{ref\_h} \right] \right) \tag{3.19}$$

$$d_3 = d_{uu} = \left(V_{ref\_g} - \left[V_{ref\_g} \right] \right) + \left(V_{ref\_h} - \left[V_{ref\_h} \right] \right) - 1$$

Following the same analysis for the case $V_3 = V_{ll}$ leads to different results for the duty cycles:

$$d_{ul} = V_{ref\_g} - \left[V_{ref\_g} \right]$$

$$d_{lu} = V_{ref\_h} - \left[V_{ref\_h} \right] \tag{3.20}$$

$$d_3 = d_{ll} = 1 - \left(V_{ref\_g} - \left[V_{ref\_g} \right] \right) - \left(V_{ref\_h} - \left[V_{ref\_h} \right] \right)$$

Equations (3.19) and (3.20) show the simplicity of duty cycle calculation: it is only necessary to round the components of the reference vector both to identify the nearest three vectors and to calculate the duty cycles. Once V_{ul}, V_{lu}, and V_3 are determined, the line voltages at the output of the converter can be calculated using the inverse transformation P^{-1} from (3.8).

However, it has to be noticed that this does not complete the switching states definition, since line voltages specify the relative voltage between phases, but nothing is said with regard to the common mode voltage with respect to the negative of the DC bus N, as explained in Figure 3.8. This degree of freedom can be exploited to address the voltage balance of the DC bus by evaluating all the leg voltage combinations (v_{aN}, v_{bN}, v_{cN}) that synthesize the line voltages dictated by V_{ul}, V_{lu}, and V_3 and selecting the one that better contributes to the voltage balance condition.

3.4 Voltage Balance Control

In order to keep the balance of the DC bus, it is necessary to evaluate how a given switching combination will modify the capacitor voltages. For this, we have to provide a method to calculate the voltage deviations of the DC bus for the different switching combinations. A method to compute the capacitor voltage deviation is presented in the next section. The following hypotheses apply:

- All capacitors have the same value ($C_1 = C_2 = \ldots = C_{n-1} = C$).
- The load is modeled as a three-wire current source.

3.4.1   Capacitor Voltage Calculation

Figure 3.21 shows the three-phase DCMC functional model where it can be seen that the position of the single-pole multiple throw switches determines each leg voltage with respect to the negative of the DC bus.

In order to keep generality, this analysis takes into account the existence or not of a DC bus voltage/power supply. This consideration lies in the fact that there are applications in which active power transfer from the AC to DC side (or vice versa) is inherent to the power processing system, for example, motor drives and active rectifiers. On the other hand, applications such as reactive power compensation and harmonics filtering do

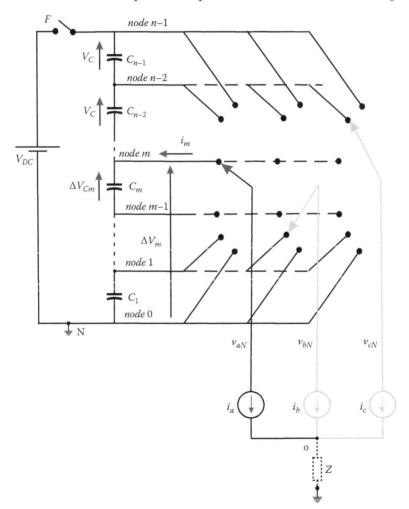

Figure 3.21 Functional model of the *n*-level three-phase DCMC converter.

not need a power source on the DC bus. Both cases are explicitly represented through the state of the switch F (Figure 3.21).

The calculation of voltage deviation in a given node of the DC bus is accomplished by individually analyzing the effects of each phase current and then summing up their contributions. This separate analysis requires a fictitious return path for the individual currents for which the dummy impedance Z between the neutral and the negative of the DC bus is included. It is clear that the effect of this impedance is globally zero provided that the neutral point is actually floating for three-wire loads and no net current flows through it.

The voltage variation in a given node of the DC bus is calculated by considering the equivalent capacitance between the node and the negative terminal of the DC bus. When $F = 1$ the current i_m flows to node m as shown in Figure 3.21. Also, i_m flows into an equivalent capacitance C_{eq_m} given by

$$C_{eq\_m} = C_{eq\_1} + C_{eq\_2} = \frac{C}{m} + \frac{C}{(n-1-m)} \tag{3.21}$$

where C_{eq_1} and C_{eq_2} are the equivalent capacitances between node m and nodes 0 and $(n-1)$, respectively. The voltage variation during the interval T_d of node m with respect to the negative of the DC bus is

$$\Delta V_m = \frac{1}{C_{eq\_m}} \int_{t_0}^{t_0+T_d} i_m(t)\,dt \tag{3.22}$$

When considering that the current is almost constant along the integration interval, the voltage increment can be simplified to

$$\Delta V_m \cong \begin{cases} = 0 & \text{if } m = 0 \text{ or } m = (n-1) \\[2ex] \dfrac{T_d\, i_m(t_0)}{C_{eq\_m}} = \dfrac{T_d\, i_m(t_0)}{\dfrac{C}{m} + \dfrac{C}{(n-1-m)}} & 1 \le m \le n-2 \end{cases} \tag{3.23}$$

The voltage deviation ΔV_m is divided among all the capacitors of the DC bus, and the polarity of the individual voltage deviation depends on the relative position of each capacitor with respect to node m, that is, if it is above or below node m. Then, the capacitor C_j ($j = 1, \ldots, n-1$) will suffer a voltage variation ΔV_{Cj}, given by

$$\Delta V_{Cj} = \begin{cases} \dfrac{\Delta V_m}{m} & 1 \le j \le m \\[2ex] \dfrac{-\Delta V_m}{(n-1-m)} & m+1 \le j \le n-1 \end{cases} \tag{3.24}$$

When the voltage source V_{DC} is not connected ($F = 0$), the equivalent capacitance between node m and node 0 is C/m, while between node m and node $n - 1$ the equivalent capacitance is zero because there is not a return path for i_m. Thus, the expression for ΔV_m is

$$\Delta V_m \cong \begin{cases} = 0 & m = 0 \\[2mm] \dfrac{T_d\, i_m\,(t_0)}{\dfrac{C}{m}} & 1 \le m \le n-1 \end{cases} \tag{3.25}$$

And the capacitor voltage variation yields to

$$\Delta V_{Cj} = \begin{cases} \dfrac{\Delta V_m}{m} & 1 \le j \le m \\[2mm] 0 & m+1 \le j \le n-1 \end{cases} \tag{3.26}$$

A generalized expression for capacitor voltage variation that explicitly takes into account both values of F can be summarized as

$$\Delta V_{Cj} \cong \begin{cases} \left(\dfrac{n-1-m}{n-1-m(1-F)}\right)\dfrac{T_d\, i_m\,(t_0)}{C} & 1 \le j \le m \\[3mm] \left(\dfrac{-Fm}{(n-1)}\right)\dfrac{T_d\, i_m\,(t_0)}{C} & m+1 \le j \le n-1 \end{cases} \tag{3.27}$$

$$F = 1: V_{DC} \text{ present}, \quad F = 0: V_{DC} \text{ absent}$$

The total voltage deviation over each capacitor is calculated summing up the contributions of the three line currents, that is, replacing i_a, i_b, and i_c instead of i_m for the corresponding values of m_a, m_b, and m_c. The voltage variation on capacitor C_j due to the three phases results in

$$\Delta V_{Cj3\phi} = \Delta V_{Cja} + \Delta V_{Cjb} + \Delta V_{Cjc} \qquad 1 \le j \le n-1 \tag{3.28}$$

3.4.2 Voltage Balance Optimization

At any sampling instant, say k, the present state of the DC bus can be represented by means of a vector whose components are the voltages across the $n - 1$ capacitors. This vector can be expressed as

$$\mathbf{V}_C[k] = \begin{bmatrix} V_{C_{n-1}}[k] & V_{C_{n-2}}[k] & \cdots\cdots & V_{C_1}[k] \end{bmatrix}^T \tag{3.29}$$

On the other hand, the reference value for a balanced DC bus is

$$\mathbf{V}_{Cref}[k] = V_{Cref}[k]\begin{bmatrix} 1 & 1 & \cdots & 1 \end{bmatrix}^T = \frac{\displaystyle\sum_{i=1}^{n-1} V_{Ci}[k]}{n-1}\begin{bmatrix} 1 & 1 & \cdots & 1 \end{bmatrix}^T \tag{3.30}$$

The voltage vector at instant $k + 1$ can be calculated in terms of the present value and the voltage variation given in (3.28):

$$\mathbf{V}_C[k+1] = \mathbf{V}_C[k] + \Delta\mathbf{V}_C[k+1] \quad \text{where:} \quad \Delta\mathbf{V}_C[k+1] = \begin{bmatrix} \Delta V_{C_{n-1}}[k+1] \\ \Delta V_{C_{n-2}}[k+1] \\ \cdot \\ \cdot \\ \Delta V_{C_1}[k+1] \end{bmatrix} \tag{3.31}$$

Also, $\mathbf{V}_C[k]$ can be written in terms of its deviation from the reference state:

$$\mathbf{V}_C[k] = \mathbf{V}_{Cref}[k] - \mathbf{E}[k] \tag{3.32}$$

Replacing (3.32) into (3.31) yields

$$\mathbf{V}_C[k+1] = \mathbf{V}_{Cref}[k] - (\mathbf{E}[k] - \Delta\mathbf{V}_C[k+1]) = \mathbf{V}_{Cref}[k] - \mathbf{E}[k+1] \tag{3.33}$$

This equation indicates the necessity of selecting the most adequate switching combination in order to minimize $\mathbf{E}[k + 1]$ and to steer the voltage vector toward the reference value $\mathbf{V}_{Cref}$. This selection can be achieved through the evaluation of a cost function that measures the difference between $\mathbf{V}_{Cref}$ and $\mathbf{V}_C$. Such a function may be the norm of $\mathbf{E}[k + 1]$:

$$\left\| \mathbf{E}[k+1] \right\|_1 = \sum_{i=1}^{n-1} \left| E_{C_i}[k] - \Delta V_{C_i}[k+1] \right| \tag{3.34}$$

From a sampling period to the next, this expression provides a way to evaluate the goodness of any converter switching combination in order to force the capacitor voltages to their reference values.

3.4.3 Flow Diagram

After the determination of the three nearest vectors $\mathbf{V}_{ul}$, $\mathbf{V}_{lu}$, and $\mathbf{V}_3$ the transformation to the line voltage sequence through the inverse transformation P^{-1} is accomplished. Also, the duty cycles set up the duration of the three line voltages such that their average within the sampling period equals $\mathbf{V}_{ref}$. Each line voltage can be realized with a set of leg voltage combinations whose maximum number of elements is five (when line voltages are zero), and the minimum number is one (when some of the three line voltages assume the DC bus voltage). Then, the sequence of line voltages that implement the reference vector can be realized, in general, with multiple sequences of leg voltages. It is worth mentioning that not all sequences are allowed since only those that imply single-step transitions on the leg voltages should be considered (Section 3.1). With this assumption, each leg combination produces different voltage deviations over the DC bus capacitors. Moreover, the total deviation of capacitor voltages along the sampling period T_S can be calculated summing up the contributions of the three leg voltage realizations:

$$\Delta V_{CjT_S} = \Delta V_{Cj(ul)} + \Delta V_{Cj(lu)} + \Delta V_{Cj(3)} \qquad j = 1, ..., n-1 \qquad (3.35)$$

where ΔV_{CjT_S} is the net voltage deviation of capacitor C_j along T_S, and $\Delta V_{Cj(ul)}$ is the voltage variation calculated with (3.28) due to the switching combination (m_a, m_b, m_c), which synthesizes $\mathbf{V}_{ul}$. Similarly, $\Delta V_{Cj(lu)}$ is the contribution to voltage deviation due to the leg voltage that synthesizes $\mathbf{V}_{lu}$, and the same applies to $\Delta V_{Cj(3)}$. Equation (3.35) is evaluated for the allowed switching sequences, and the one that minimizes (3.34) is selected to synthesize the set $\mathbf{V}_{ul}$-$\mathbf{V}_{lu}$-$\mathbf{V}_3$. Figure 3.22 shows the flow diagram of the entire modulation process.

3.5 Effectiveness Boundary of Voltage Balancing in DCMC Converters

Redundant states are the key for capacitor voltage balancing control when additional hardware is not considered. However, although it is possible to extend the operation of the converter to active power processing, this approach still has limitations. In particular, n-level three-phase converters with $n > 3$ cannot maintain a balancing condition when supplying current with a high power factor at high modulation index, even when optimum selection of switching states is carried out. Moreover, the result presented by Marchesoni and Tenca [15] states that for a large number of levels ($n \to \infty$), there is a bidimensional domain of modulation index (M) and load

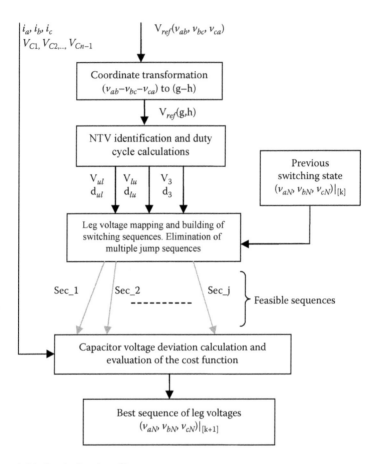

Figure 3.22 Optimization diagram.

power factor (cos φ). Inside this domain, there is no modulation strategy that is able to simultaneously synthesize a sinusoidal voltage waveform and preserve the voltage balance of the DC bus capacitors. This analysis yields to Equation (3.36), which describes the region within which there is no balancing possibility of the three-phase DCMC in the terms described above:

$$M\cos(\varphi) > \frac{\sqrt{3}}{\pi} \approx 0.55 \qquad (3.36)$$

The demonstration of (3.36) is supported on a balancing theorem introduced by Marchesoni et al. [8], which states the general conditions of any modulation algorithm to keep the voltage balance of the DC bus. The

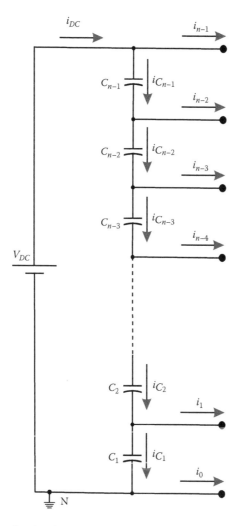

Figure 3.23 Figure for the theorem proof.

theorem applies to finite or infinite number of levels, and it states, based on Figure 3.23, three necessary and sufficient conditions to ensure the balance (the quantities are average values). For an n-level DCMC:

$$(a) \ \langle i_m \rangle = 0 \ \text{ for } \ m = 1, ..., n-2$$

$$(b) \ \langle i_{n-1} \rangle = \langle i_{DC} \rangle \tag{3.37}$$

$$(c) \ \langle i_0 \rangle = -\langle i_{DC} \rangle$$

The proof of this theorem begins considering the generic equation for the current through capacitor C_j:

$$i_{C_j} = i_{DC} - \sum_{m=j}^{n-1} i_m \quad \text{for } j = 1, \dots, n-1 \tag{3.38}$$

Considering the average values of both sides of (3.38),

$$\langle i_{C_j} \rangle = \langle i_{DC} \rangle - \sum_{m=j}^{n-1} \langle i_m \rangle \quad \text{for } j = 1, \dots, n-1 \tag{3.39}$$

Considering also that the average current through all capacitors should be zero for null net voltage variation,

$$\begin{vmatrix} \langle i_{C_{n-1}} \rangle = \langle i_{DC} \rangle - \langle i_{n-1} \rangle = 0 \\ \langle i_{C_{n-2}} \rangle = \langle i_{DC} \rangle - \langle i_{n-2} \rangle - \langle i_{n-1} \rangle = 0 \\ \langle i_{C_{n-3}} \rangle = \langle i_{DC} \rangle - \langle i_{n-3} \rangle - \langle i_{n-2} \rangle - \langle i_{n-1} \rangle = 0 \\ \dots \\ \langle i_{C_1} \rangle = \langle i_{DC} \rangle - \langle i_1 \rangle - \dots - \langle i_{n-3} \rangle - \langle i_{n-2} \rangle - \langle i_{n-1} \rangle = 0 \end{vmatrix} \quad \text{leads to:} \begin{cases} \langle i_{n-1} \rangle = \langle i_{DC} \rangle \\ \langle i_{n-2} \rangle = 0 \\ \langle i_{n-3} \rangle = 0 \\ \dots \\ \langle i_1 \rangle = 0 \end{cases} \tag{3.40}$$

Moreover, as all the currents sum zero,

$$\sum_{m=0}^{n-1} i_m = 0 \quad \text{implies:} \quad \langle i_0 \rangle = -\langle i_{n-1} \rangle = -\langle i_{DC} \rangle \tag{3.41}$$

This result states that all the power delivered to the load has to be transferred through the most extreme taps of the DC bus, and that the modulation strategy should manage the redundant states of the converter in order to balance the incoming and outgoing currents on the intermediate taps. This idea was given intuitively in Figure 3.6.

From this result, we can work over (3.37) to search a boundary of validity for conditions (b) and (c). For simplicity, we analyze the case of a single-phase load and then extend the reasoning to a three-phase case.

Figure 3.24 shows an n-level DC bus and a single-phase load whose movable taps A and B can be arbitrarily connected to the fixed taps 0

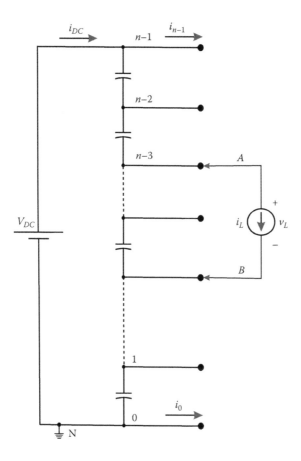

Figure 3.24 Single-phase load connected to an n-level capacitor divider.

to $n - 1$. In addition, two hypotheses are considered for the following reasoning:

1. The number of levels n is infinite; that is, the load slides continuously through the fixed taps 0 to $n - 1$, and consequently the voltage waveform v_L is continuous.
2. The synthesized voltage v_L and the current i_L are sinusoidal with a relative phase displacement φ ($-\pi < \varphi < \pi$).

Based on (2), we can write:

$$v_L(\omega t) = V_p \sin(\omega t) = MV_{DC} \sin(\omega t) \quad \text{and:} \quad i_L(\omega t) = I_p \sin(\omega t - \varphi) \quad (3.42)$$

In steady-state operation, the average power delivered or absorbed by the DC source must be equal to the power at the AC side. Then, the average current at the DC side can be determined:

$$P_{DC} = V_{DC}\langle i_{DC}\rangle = P_{AC} = V_{RMS}I_{RMS}\cos(\varphi) = \frac{MV_{DC}}{\sqrt{2}}\frac{I_p}{\sqrt{2}}\cos(\varphi) \Rightarrow$$

(3.43)

$$\Rightarrow \langle i_{DC}\rangle = \frac{MI_p}{2}\cos(\varphi)$$

On the other hand, the balancing conditions (b) and (c) from (3.37) require that the average value $\langle i_{n-1}\rangle$ be equal to $-\langle i_0\rangle$, and this in turn depends on the capability of the load current i_L to inject balanced average currents into these taps. In this sense, considering the sinusoidal shape for i_L, the maximum value of $\langle i_{n-1}\rangle$ and $-\langle i_0\rangle$ is obtained draining or injecting the positive half cycle of the sinusoid in node $n-1$ and the negative half cycle in node 0. This statement is supported by the fact that the movable taps A and B are never connected simultaneously to the taps 0 and $n-1$ during a finite time interval, even for $M = 1$ (in fact, the sinusoidal waveform does imply a simultaneous connection between taps 0 and $n-1$, but this state does not last more than 0 seconds, when a peak voltage $v_{L\ peak} = V_{DC}$ is synthesized). In these conditions, the value of the maximum balanced average current that can be injected into nodes 0 and $n-1$ is

$$\langle i_{n-1}\rangle_{max} = -\langle i_0\rangle_{max} = \frac{1}{2\pi}\int_0^{\pi} i_L(\omega t)d\omega t = \frac{I_p}{\pi}$$

(3.44)

This value is independent of M and φ, and for sure the average current delivered by the voltage source V_{DC} cannot be higher than (3.44). This is expressed in (3.45):

$$\langle i_{DC}\rangle \leq \langle i_{n-1}\rangle_{max}$$

(3.45)

Replacing (3.43) and (3.44) into (3.45), we have

$$\frac{MI_p}{2}\cos(\varphi) \leq \frac{I_p}{\pi} \quad \Rightarrow \quad M\cos(\varphi) \leq \frac{2}{\pi} \approx 0.63$$

(3.46)

Equations (3.43) and (3.46) ensure the compliance of conditions (b) and (c) of the balancing theorem. It is clear that the compliance of (3.46) does not self-imply voltage balance, since nothing was taken into account regarding zero-average current on the internal nodes of the bus. Nevertheless, a general result can be stated in the following terms: there is no balancing strategy for the single-phase infinite levels DCMC if the condition (3.47) is satisfied:

$$M\cos(\varphi) > \frac{2}{\pi} \tag{3.47}$$

This reasoning can be extended to the three-phase DCMC, taking into account the following points:

- Three balanced sinusoidal line voltages are synthesized at the AC side of the converter.
- The three line currents are sinusoidal and balanced with a phase shift equal to φ.

In this case, the average current at the DC side is calculated recalling the power balance between the AC and DC sides:

$$
\begin{aligned}
P_{DC} = V_{DC}\langle i_{DC}\rangle &= P_{AC} = 3V_{phaseRMS} I_{phaseRMS} \cos(\varphi) \\
&= 3\left(\frac{MV_{DC}}{\sqrt{3}\sqrt{2}}\right)\left(\frac{I_p}{\sqrt{2}}\right)\cos(\varphi) \\
\Rightarrow \quad \langle i_{DC}\rangle &= \frac{\sqrt{3}MI_p}{2}\cos(\varphi)
\end{aligned}
\tag{3.48}
$$

Under the same considerations explained for the single-phase case, the maximum average current that can be injected to the fixed tap $n-1$ by the three phase currents (considering $\langle i_{n-1}\rangle = -\langle i_0\rangle$) is the average value of the six-pulse rectified waveform divided by 2:

$$\langle i_{n-1}\rangle_{max} = -\langle i_0\rangle_{max} = \frac{1}{2}\frac{\sin(\pi/6)}{\pi/6}I_p \tag{3.49}$$

Again, the current at the DC side cannot be higher than (3.49). Then:

$$\langle i_{DC}\rangle \le \langle i_{n-1}\rangle_{max} \Rightarrow M\cos(\varphi) \le \frac{\sin(\pi/6)}{\sqrt{3}\pi/6} = \frac{\sqrt{3}}{\pi} \tag{3.50}$$

And for the three-phase infinite-level DCMC it can be assured that no balancing strategy exists if the condition (3.51) is satisfied:

$$M\cos(\varphi) \; > \; \frac{\sqrt{3}}{\pi} \approx 0.55 \qquad\qquad (3.51)$$

The same considerations given for (3.47) apply to (3.51), in the sense that the pairs (M, φ) that comply with (3.51) necessarily imply voltage unbalance of the DC bus. Nevertheless, the nonaccomplishment of (3.51) does not ensure voltage balance unless the converter redundant states are properly managed. Moreover, additional restrictions of redundant states selection, such as converter losses or single-jump conditions, may expand the unbalance boundary, but this depends on the particular modulation scheme.

The conditions (3.47) and (3.51) refer to an infinite-level DC bus, which allows the assumption of sinusoidal voltages with zero time interval connection between taps 0 and $n-1$, even at maximum modulation index. When the number of levels is small, however, the voltage waveform is far from being a perfect sinusoid, thus relaxing the hypothesis of non-simultaneous connection of the load currents between extreme taps. As a consequence, the division factor 2 that was considered for the balance of the average currents on taps 0 and $n-1$ is reduced. This reduction depends on the average time that the load holds the connection between the taps 0 and $n-1$, and the longer this time is, the factor 2 is reduced and the operation boundary may be expanded. Again, this factor depends on the adopted modulation strategy, and its precise determination requires a detailed analysis of the switching patterns.

3.6 Performance Results

The control algorithms are tested by means of computer simulations. A five-level DCMC is connected to a three-phase distribution grid through a coupling inductor as shown in Figure 3.25. The DC bus voltage is set to 20 kV, while the coupling inductor is 5 mH, $C_{bus} = 4700\ \mu F$, and the averaging period T_S is set to 0.4 ms. The power system is rated to a line voltage of 14 kV.

The reference vector $\mathbf{V}_{ref}$ is synchronized to the main voltage, and the current flowing through the coupling inductor is set by means of the amplitude of the reference voltage vector.

Figure 3.26a shows one line voltage at the converter terminals. As it can be seen, it exhibits single-step transitions over the complete trace featuring good harmonic performance. The currents provided by the converter to the coupling inductor are shown in Figure 3.26b. The three line currents that circulate through the coupling inductor have a large component

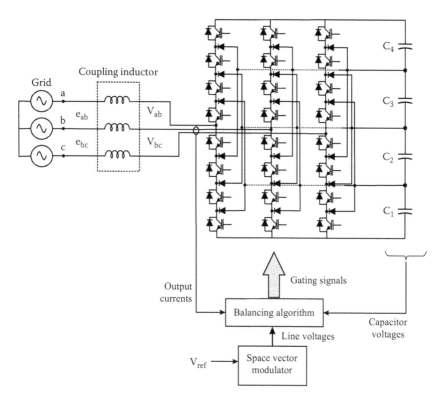

Figure 3.25 Five-level DCMC connected to the power grid through a coupling inductor.

at the fundamental frequency with a small switching frequency ripple. Figure 3.27 shows the capacitor voltages that remain balanced at their reference value of 5 kV, with an approximate ripple of 200 V peak to peak.

In order to test the balancing algorithm, a voltage unbalance is forced externally to the DC bus capacitors. With a modulation index of $M = 0.85$, the unbalance is set at $t = 100$ ms. The capacitor voltages are shown in Figure 3.28. The unbalancing condition is mitigated and the balance is restored in approximately 100 ms. Figure 3.29 shows single jumps on leg voltages, even at the disturbance instant.

Finally, the modulation index is modified to $M = 0.96$ and the same disturbance on capacitor voltages is introduced. It is observed that the required time for balancing restoration is slightly extended (Figure 3.30) compared with the previous case. On the other hand, fewer transitions are observed on the leg voltages (Figure 3.31). This makes sense since the operation at higher modulation index implies less redundant states, which limits the balancing action on capacitor voltages. This also indicates that it is not possible to operate in the overmodulation region since

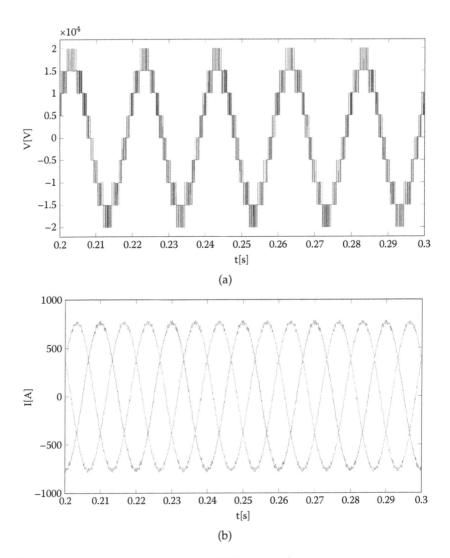

Figure 3.26 (a) Converter line voltage. (b) Converter line currents.

in this case no redundant states are available and the balancing action becomes impossible.

3.7 Summary

In this chapter, an analysis of the diode-clamped multilevel converter has been performed. The working principle of the converter leg has been explained. The set of usable states is resumed and a proper logic gating for the topology was synthesized. Also, the transient switching characteristic

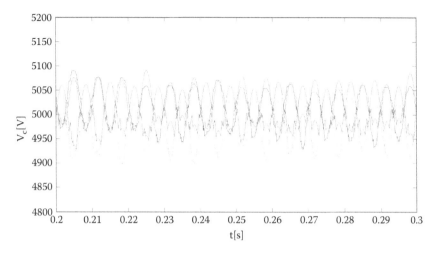

Figure 3.27 DC capacitor voltages.

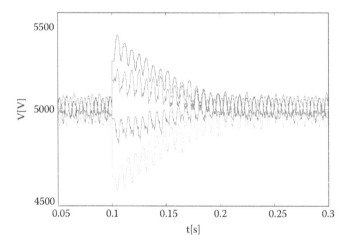

Figure 3.28 Capacitor voltages with forced unbalancing ($M = 0.85$).

is analyzed and the single-jump restriction for leg voltages is concluded as a general rule for the transition between consecutive states. Also, the line voltage redundancy is introduced and the legs combinations are analyzed in terms of their effects on capacitors' voltages. Using these concepts, a generalized algorithm for DC bus voltage balance of the diode-clamped multilevel converter has been described and a multilevel space vector modulator was explained. The SVM was used jointly with the balancing strategy in order to synthesize the desired line voltages at the output maintaining the voltage balance of the DC bus for safe operation of the power devices.

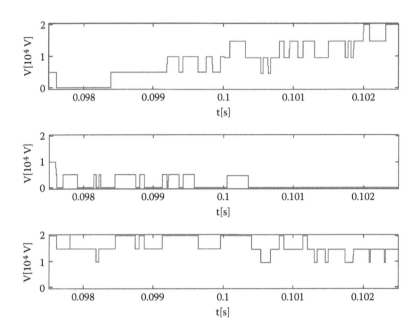

Figure 3.29 Leg voltages at unbalancing instant (*M* = 0.85).

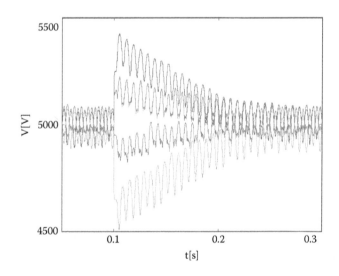

Figure 3.30 Capacitor voltages with forced unbalancing (*M* = 0.96).

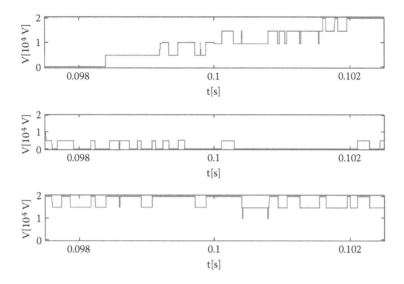

Figure 3.31 Leg voltages at unbalancing instant ($M = 0.96$).

References

1. A. Nabae, I. Takahashi, H. Akagi. A New Neutral-Point-Clamped PWM Inverter. *IEEE Transactions on Industry Applications*, 17(5), 518–523, 1981.
2. T.A. Meynard, H. Foch. Multi-Level Choppers for High Voltage Applications. *European Power Electronics and Drives Journal*, 2(1), 45–50, 1992.
3. X. Yuan, I. Barbi. Fundamentals of a New Diode Clamping Multilevel Inverter. *IEEE Transactions on Power Electronics*, 15(4), 711–718, 2000.
4. P. Bartolomeüs, P. Le Moigne, J.B. Mba. Over-Voltage Problems of Diode-Clamped Converters during Switching. In *European Conference on Power Electronics and Applications (EPE'03)*, Toulouse, France, September 2–4, 2003.
5. G.P. Adams, S.J. Finney, A.M. Massoud, B.W. Williams. Capacitor Balance Issues of the Diode Clamped Multilevel Inverter Operated in Quasi-Two State Mode. *IEEE Transactions on Industrial Electronics*, 55(8), 3088–3099, 2008.
6. G.P. Adams, S.J. Finney, A.M. Massoud, B.W. Williams. Two Level Operation of a Diode Clamped Multilevel Inverter. In *IEEE International Symposium on Industrial Electronics (ISIE 2010)*, Bari, Italy, July 4–7, 2010, pp. 1137–1142.
7. C. Hochgraf, R. Lasseter, D. Divan, T. Lipo. Comparison of Multilevel Converters for Static VAR Compensation. In *IEEE Industry Applications Society Annual Meeting (IAS'94)*, Denver, CO, October 2–6, 1994, pp. 921–928.
8. M. Marchesoni, M. Mazzucchelli, P. Tenca. About the DC-Link Capacitors Voltage Balance in Multi-Point Clamped Converters. In *IEEE Annual Conference of the Industrial Electronics Society (IECON'98)*, Aachen, Germany, August 31–September 4, 1998, pp. 548–553.
9. K.A. Corzine, J. Yuen, J.R. Baker. Analysis of a Four-Level DC/DC Buck Converter. *IEEE Transactions on Industrial Electronics*, 49(4), 746–751, 2002.

10. N. Hatti, K. Hasegawa, H. Akagi. A 6.6-kV Transformerless Motor Drive Using a Five-Level Diode-Clamped PWM Inverter for Energy Savings of Pumps and Blowers. *IEEE Transactions on Power Electronics*, 24(3), 796–803, 2009.

11. D.G. Holmes, T.A. Lipo. *Pulse Width Modulation for Power Converters: Principles and Practice*. IEEE Press, Piscataway, NJ, 2003.

12. J.-H. Suh, C.-H. Choi, D.-S. Hyun. A New Simplified Space-Vector PWM Method for Three-Level Inverters. *IEEE Transactions on Power Electronics*, 16(4), 545–550, 2001.

13. R. Joetten, C. Kehl. A Fast Space Vector Control for a Three Level Voltage Source Inverter. In *European Conference on Power Electronic and Applications (EPE'91)*, Firenze, Italy, September 1991, pp. 70–75.

14. N. Celanovic, D. Boroyevich. A Fast Space-Vector Modulation Algorithm for Multilevel Three-Phase Converters. *IEEE Transactions on Industry Applications*, 37(2), 637–641, 2001.

15. M. Marchesoni, P. Tenca. Theoretical and Practical Limits in Multilevel MPC Inverters with Passive Front Ends. In *Conference on Power Electronic and Applications (EPE'01)*, Graz, Austria, August 27–29, 2001.

chapter 4

Flying Capacitor Multilevel Converter

4.1 Introduction

The flying capacitor multilevel converter (FCMC) has been introduced by Meynard and Foch [1]. This topology was conceived to implement high-voltage converters without series connection of the power switches. The FCMC was characterized, in Chapter 2, as a symmetric topology. Also, it was possible to see that the FCMC has some advantages when compared to the diode-clamped multilevel converter (DCMC). It is easy to increase the number of voltage levels by simply adding basic cells in the load end of each leg. The same as the DCMC, it has a single voltage supply that constitutes the DC bus, allowing the back-to-back connection of these converters. As with the DCMC, the FCMC requires some strategy to maintain the voltage balance on the different capacitors, but in this case, this is easily overcome using the redundant states in each leg of the converter with an adequate modulation strategy. The main disadvantage of this topology is that it requires a large number of capacitors since it builds the voltage levels with flying capacitors in each leg of the converter.

The output voltage of a FCMC is generated through different connections of the flying capacitors. It is very important that all the FCs reach a constant and stable voltage, so the net charge variation on each of them should be null. There are two important reasons for this. The first is to reduce the harmonic distortion on the output voltage. The second is to guarantee the same blocking voltage of each power switch equals the same fraction of the total DC bus voltage.

In this chapter, the FCMC topology is first described. Then, a phase-shifted carrier pulse width modulation (PSPWM) that allows preserving the voltage balance is introduced [2]. Although this addresses the FC voltage stabilization in steady state, some different action is required for the initial charge of the flying capacitors and to correct transient deviations [3]. The simplest way to deal with this problem is to inject an equalizing current through an additional passive network.

4.2 Flying Capacitor Topology

It was shown in Chapter 2 that each flying capacitor belongs to an intermediate basic cell of the generalized topology, where all the inner switches
were eliminated. The voltage across each capacitor (V_C) preserves the same
value as in the generalized topology. For an n-level converter,

$$V_C = \frac{V_{DC}}{n-1} \qquad (4.1)$$

An n-level converter employs $n - 1$ stages, where each stage is controlled with a switching function (s_j). The jth stage of a FCMC has a pair
of complementary switches plus ($n - j$) capacitors connected in series. The
combined action of the switching functions of each stage generates the leg
voltage v_{iN}, given by

$$v_{iN} = \frac{V_{DC}}{(n-1)} \cdot \sum_{j=1}^{n-1} s_j \qquad (4.2)$$

An n-level FCMC requires $(n - 1)(n - 2)/2$ flying capacitors in each leg
of the converter plus $(n - 1)$ capacitors on the DC bus. This number can be
reduced if the flying capacitors withstand different voltages. Then, the ($n
- j$) capacitors of each stage may be replaced by a single capacitor, and each
leg has only $(n - 2)$ capacitors. Now each capacitor works with an average
voltage equal to

$$V_j = j\frac{V_{DC}}{n-1} \qquad (4.3)$$

where $j = 1$ corresponds to the stage beside the load.

Figure 4.1 shows one leg of a four-level converter that employs one
capacitor per stage. The DC bus, common to all legs of the converter, may
be implemented by a voltage DC source or by a single capacitor. In this
example, the average voltages over the flying capacitors are $2/3V_{DC}$ and
$1/3V_{DC}$ for the second and third stages, respectively.

4.2.1 Charge Balance on the Flying Capacitors

It was analyzed in Chapter 2 that it was necessary to use all the redundant
states to maintain the average voltage on each flying capacitor. This condition implies that the overcharge or undercharge in one state is eliminated
or regained in the redundant state.

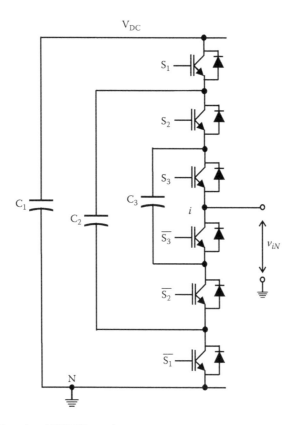

Figure 4.1 Four-level FCMC topology.

Figure 4.2 shows a general stage of a FCMC together with the power switches of the previous stage. This figure is useful to analyze the voltage balance on the flying capacitor. In order to maintain constant the average voltage on C_{ji}, it is necessary that the net charge (ΔQ_{Cji}) along one cycle should be null. The following condition should be accomplished along a switching cycle T_S, in steady state:

$$\Delta Q_{Cji} = \int_{Ts} i_{Cji} \cdot dt = 0 \tag{4.4}$$

Then, it is necessary to know the current i_{Cji} for each switching function of the related power switches.

Looking to Figure 4.2, the current i_{Cji} can be expressed as the difference of the currents through the switches $S_{(j-1)i}$ and S_{ji}, $i_{S(j-1)i}$ and i_{Sji}, respectively. Since the power switches carry the load current (i_i), their currents can be written as

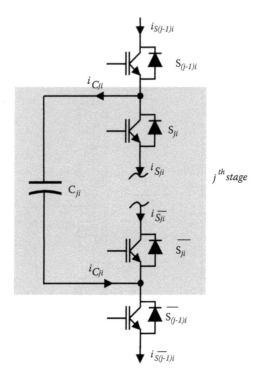

Figure 4.2 One general stage of the FCMC.

$$i_{S(j-1)i} = s_{(j-1)i}i_i$$

$$i_{Sji} = s_{ji}i_i$$

(4.5)

Then, the current across C_{ji} is

$$i_{Cji} = (s_{(j-1)i} - s_{ji}) \cdot i_i$$ (4.6)

Equation (4.6) gives two possible conditions to meet the charge balance given by (4.4). The first condition should be analyzed when the power switches operate at line frequency. In this case the load current (i_i) changes its sign along T_S; therefore, i_{Cji} also changes its sign, so (4.4) is satisfied. The second possibility is when the switches operate at high frequency and the load current may be considered as constant along T_S. In this case, (4.4) is satisfied with the corresponding switching functions in (4.6). This requires the use of the redundant states of the FCMC.

4.3 Modulation Scheme for the FCMC

Different modulation schemes for the FCMC can be found in the bibliography, such as space vector modulation [4], predictive control [5], selective harmonics elimination [6], or carrier-based pulse width modulation with several carriers [7]. Any of these modulations should satisfy (4.4) in order to guarantee the charge balance of the flying capacitors. Among them, the phase-shifted carrier pulse width modulation (PSPWM) [2] naturally provides the required charge and voltage balance of all the capacitors of the FCMC. This modulation travels along the redundant states of the FCMC providing self-charge balancing without additional controllers.

4.3.1 Phase-Shifted Carrier Pulse Width Modulation

The PSPWM of an n-level FCMC requires $n-1$ carriers with a phase displacement as shown in Figure 4.3. Each carrier is a triangular wave with an amplitude A_p and a frequency f_s much higher than the modulating frequency. The phase shift among them is equal to

$$\delta = \frac{2\pi}{n-1} \tag{4.7}$$

A single modulating signal is compared with each carrier generating $n-1$ switching functions. The modulation index (m) is defined as the ratio between the amplitudes of the modulating signal and the carrier,

$$m = \frac{V_m}{A_p} \tag{4.8}$$

While m is lower than unity, there exists a linear relationship between the amplitude of the fundamental component of the leg voltage (v_{iN}) and V_{DC}.

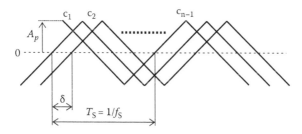

Figure 4.3 Carriers in PSPWM.

The power devices of the FCMC switch at a frequency f_S, while the leg voltage (v_{iN}) commutates at a frequency equal to $(n - 1).f_S$. Then, the harmonics spectrum of the load voltage is shifted to higher frequencies, simplifying the design of the output filter.

4.3.1.1 Charge Balance Using PSPWM

The PSPWM in an n-level FCMC naturally maintains the charge balance in all the flying capacitors. This is done traveling along the redundant states in each switching cycle. This is easily analyzed for the three-level FCMC, shown in Figure 4.4. The same analysis procedure can be extended to an n-level converter. The three-level FCMC converter requires two carriers with a phase shift equal to π. Figure 4.5a,b shows several cycles of both carrier waveforms, c_1 and c_2, and a fraction of the modulating signal (positive in Figure 4.5a and negative in Figure 4.5b). The comparisons between the carrier waveforms and the modulating signal generate the switching functions (s_{1i} and s_{2i}) that drive the power switches S_{1i} and S_{2i}, respectively. Taking into account (4.2) for $n = 3$, when the modulating signal is positive, the output voltage changes between $V_{DC}/2$ and V_{DC}. On the other hand, when the modulating signal is negative, v_{iN} changes between 0 and $V_{DC}/2$.

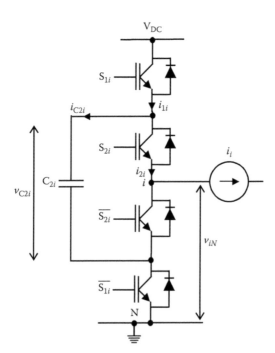

Figure 4.4 One leg of a three-level FCMC.

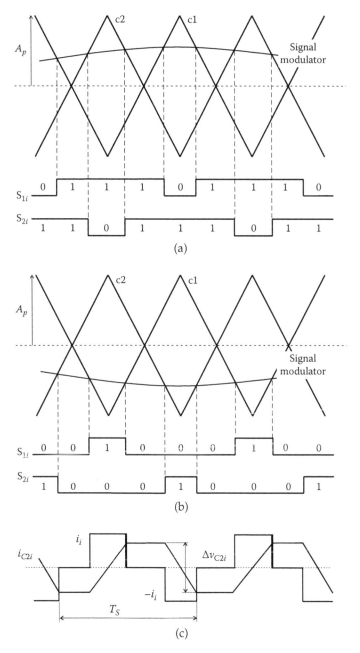

Figure 4.5 PSPWM. Switching functions (s_{1i} and s_{2i}) for $n = 3$. (a) Switching functions when the modulating signal is positive, (b) switching functions when the modulating signal is negative, and (c) current and voltage on C_{2i}.

The analysis of the voltage balance on the flying capacitor of Figure 4.4 is done looking to the current and voltage on C_{2i}. Figure 4.5c shows the voltage variation of C_{2i} around its average value along one switching cycle. It is valid for either positive or negative modulation signals.

The study begins with state 01, which is valid for either sign of the modulating signal. In this state the switches $\bar{S}_{1i}$ and S_{2i} are ON, and Figure 4.4 shows that $i_{C2i} = -i_i$. The capacitor C_{2i} loses some charge making a voltage drop equal to Δv_{C2i}. In either state 00 or 11 the current i_{C2i} is null; then the voltage v_{C2i} remains constant at the level gained in the previous state. When state 10 is activated, S_{1i} and $\bar{S}_{2i}$ are ON; then $i_{C2i} = i_i$ and the capacitor regains the charge lost while being in state 01. In this way, the voltage v_{C2i} increases an amount Δv_{C2i}, equal to the decrement, while in 01 if the time intervals that the converter remains in both states are the same then the average voltage along T_S equals $V_{DC}/2$. The time that the converter remains in each state depends on the modulating signal. When the carrier frequency is much higher than the modulating frequency, the current may be considered constant during a carrier cycle. Then the time intervals that the converter remains in each redundant state are almost identical.

Figure 4.6 shows the switching functions and the leg voltage along the whole cycle of the reference signal. In this case the carrier frequency is 20 times higher than the reference signal of 50 Hz. The leg voltage jumps between $V_{DC}/2$ and V_{DC} during the positive semicycle of the reference. Otherwise, it jumps between $V_{DC}/2$ and 0 along the negative semicycle. Since

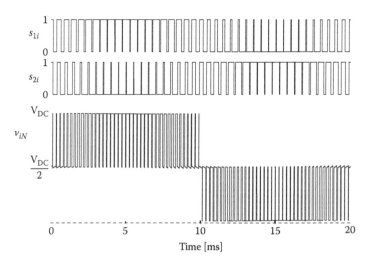

Figure 4.6 Switching function and voltage waveforms with PSPWM along one modulation cycle.

the load current i_i is a 50 Hz sinusoid, the intermediate voltage level ($V_{DC}/2$) maintains its average value along the whole cycle of the reference signal.

4.4   Dynamic Voltage Balance of the FCMC

In the previous section it was shown, for a three-level FCMC, that the PSPWM maintains the charge balance in steady state. This is true assuming that the average voltage of C_{2i} has already reached the value $V_{DC}/2$. But this is not the only stable condition. It may happen that the voltage v_{C2i} has a steady-state value different from the expected $V_{DC}/2$. If this happens, the leg voltage v_{iN} will have asymmetric voltage levels, leading to a loss in waveform symmetry. This causes several drawbacks: voltage jumps of different amplitude, asymmetric voltage waveform, increased harmonics spectrum, and increased blocking voltage for the power switches of the converter. Then, it is very important to find a strategy in order to guarantee that every flying capacitor reaches the desired voltage.

The dynamic behavior of the voltage in the flying capacitors has been treated in the literature through the harmonics components [8,9]. It has been demonstrated that the voltage unbalances on the flying capacitors of an n-level FCMC can be corrected introducing harmonic currents at frequencies equal to $k.f_s$ (where k is an integer but not a multiple of the number of levels n) [10]. There exist two mechanisms through which these harmonics can appear on the flying capacitors. The first one deals with the load current as was analyzed by Yuan et al. [9] and McGrath and Holmes [11]. In this way, the time of convergence to the steady state depends on the load. The other method consists in generating the balancing harmonics with passive networks tuned at the switching frequency. This network not only allows the dynamic balance of flying capacitors, it can also fix the convergence time to the steady state, through a proper design of the passive network.

4.4.1   Dynamic Model

It is better to use a time domain model to analyze the dynamic behavior of the currents through the flying capacitors [12]. The circuit presented in Figure 4.4 is used, assuming that the switching frequency is much higher than the modulating frequency; in this case, it is valid to assume that the duty cycles of both switches are equal, since the current is almost constant along one commutation period.

The increment of the averaged voltage on capacitor C_{2i} is much slower than the voltage ripple generated by the switching frequency [13]. These averaged voltage variations are denoted with the symbol ~. A variation of the averaged voltage across the capacitor implies that there is an average current through it. Both are related by

$$\tilde{i}_{C2i} = C_{2i} \frac{d\tilde{v}_{C2i}}{dt} \tag{4.9}$$

where $\tilde{i}_{C2i}$ is the variation of the averaged current within a commutation cycle (T_S). Recalling (4.6),

$$i_{C2i} = (s_{1i} - s_{2i}) \cdot i_i \tag{4.10}$$

Assuming that i_i is almost constant along T_S, the averaged current may be calculated as

$$\tilde{i}_{C2i} = \frac{i_i}{T_S} \int_{T_S} (s_{1i} - s_{2i}) \cdot dt \tag{4.11}$$

In (4.11) the averages of the switching functions are the duty cycles d_{1i} and d_{2i} of the power switches S_{1i} and S_{2i}, respectively. Replacing (4.11) in (4.9) results in

$$\frac{d\tilde{v}_{C2i}}{dt} = \frac{1}{C_{2i}} (d_{1i} - d_{2i}) i_i \tag{4.12}$$

It is easy to see in (4.12) that, for a given load current, the voltage $\tilde{v}_{C2i}$ is controlled with the duty cycles d_{1i} and d_{2i}. This requires a voltage control loop for the flying capacitor. When using a PSPWM, both duty cycles are equal; so regardless of load current i_i, the voltage v_{C2i} will remain in its steady-state value.

Assuming that the voltage ripple is much smaller than its average value, the output leg voltage v_{iN} equals the sum of the voltages across the switches $\bar{S}_{1i}$ and $\bar{S}_{2i}$. Then,

$$v_{iN} = (\tilde{v}_{DC} - \tilde{v}_{C2i}) s_{1i} + \tilde{v}_{C2i} s_{2i} \tag{4.13}$$

Here, it is assumed that the DC bus voltage ($\tilde{v}_{DC}$) may also vary.

The averaged voltage of the leg is calculated from (4.13), assuming slow voltage variations. Then,

$$\tilde{v}_{iN} = (\tilde{v}_{DC} - \tilde{v}_{C2i}) \cdot d_{1i} + \tilde{v}_{C2i} \cdot d_{2i} \tag{4.14}$$

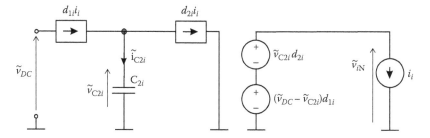

Figure 4.7 Averaged model of a three-level FCMC.

Equations (4.12) and (4.14) describe the averaged model of a three-level FCMC, which is represented by the circuit shown in Figure 4.7. The output circuit is built with a current generator, representing the load, connected in series with two voltage generators that depend on the DC and capacitor voltages and the duty cycles as well. The input circuit is modeled with two current generators that depend on i_i, d_{1i}, and d_{2i}, and are connected to C_{2i}. It is clearly seen in this circuit that any difference between the duty cycles of both switches will modify the current generators, resulting in a voltage variation on the capacitor.

During start-up, the flying capacitor is initially discharged. So it is necessary to introduce a difference in the duty cycles in order to reach the steady-state value. In this case the phase-shifted carrier modulation is not capable of generating the charge of the flying capacitor. So it is convenient to find a different way to charge the capacitor so that $\tilde{v}_{C2i}$ reaches its steady-state value independently of the duty cycles of the power switches and the load current. The idea is to introduce another current in (4.12) that is present only during transients. This additional current can be added using a tuned network to perform the balancing action, as shown in Figure 4.8. The behavior of this network is analyzed in the next section.

4.4.2   Tuned Balancing Network

Looking to (4.13), it is clear that if $\tilde{v}_{C2i}$ is not exactly equal to $V_{DC}/2$, the leg voltage pulse (v_{iN}) adopts four different values along a commutation cycle $(0, \tilde{v}_{DC} - \tilde{v}_{C2i}, \tilde{v}_{C2i}, \tilde{v}_{DC})$. This operating condition is illustrated in Figure 4.9a for one switching cycle when the modulating signal is negative. In this case the voltage v_{C2i} is lower than $V_{DC}/2$. Then, from (4.13), v_{iN} is higher than $V_{DC}/2$ when S_{1i} is ON, and it is lower than $V_{DC}/2$ when S_{2i} is ON. Then v_{iN} presents an amplitude modulation instead of being a square wave with amplitude equal to $V_{DC}/2$. The leg voltage can be divided into two components: a transient one (v_{it}) that is an alternate waveform with

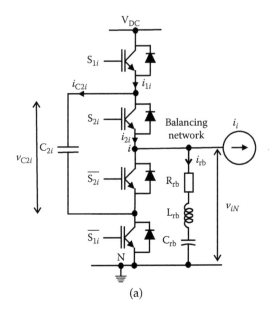

(a)

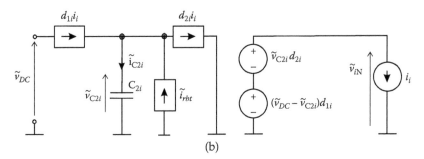

(b)

Figure 4.8 (a) Three-level FCMC with balancing network. (b) Averaged model of FCMC with balancing network.

frequency equal to the switching frequency, and a square waveform (v_{iss}) that corresponds to the steady-state case as shown in Figure 4.9b.

When connecting a balancing network ($R_{rb} - L_{rb} - C_{rb}$), tuned at the switching frequency, to the output of the leg (Figure 4.8a), an alternating current will circulate through it. Both, the v_{iss} and v_{it} voltages, lead to different currents on the balancing network, i_{rbss} and i_{rbt}, respectively. The i_{rbt} current together with the switches S_{1i} and S_{2i} will generate an averaged current ($\tilde{i}_{rbt}$) through the flying capacitor C_{2i}. This current will charge or discharge the flying capacitor, so that its voltage converges to $V_{DC}/2$.

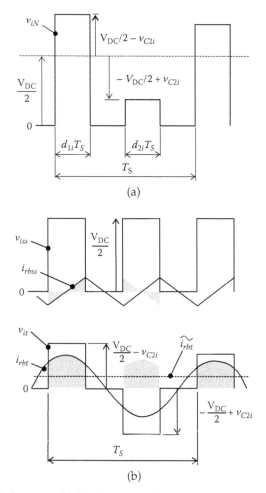

Figure 4.9 (a) Voltage v_{iN}. (b) Steady-state and transient components of v_{iN} and i_{rb}.

When $\tilde{v}_{C2i}$ reaches its steady state, v_{iN} is a square wave with a frequency equal to the double of the switching frequency and $\tilde{i}_{rbt}$ tends to zero.

If the balancing network is tuned at the switching frequency, and if it has a high merit factor (Q_{rb}), the transient current i_{rbt} is almost sinusoidal. On the other hand, at steady state (double the switching frequency) the balancing network is highly inductive and the current i_{rbss} is a triangular waveform of small amplitude. Then, the current on the balancing network can be expressed as the sum of both components:

$$i_{rb} = i_{rbss} + i_{rbt} \tag{4.15}$$

According to (4.10), the current through C_{2i} has a new value,

$$i_{C2i} = \left(s_1 - s_2\right)\left(i_i + i_{rb}\right) \tag{4.16}$$

While i_i is almost constant along a switching cycle, i_{rb} changes its sign, as can be observed in the transient component in Figure 4.9b. When $s_1 = 1$ and $s_2 = 0$, i_{C2i} is calculated as

$$i_{C2i} = s_1 \left(i_i + i_{rbss} + i_{rbt}\right) \tag{4.17}$$

On the other hand, when $s_1 = 0$, $s_2 = 1$, and considering the change of polarity of i_{rbt}, i_{C2i} results in

$$i_{C2i} = s_2 \left(-i_i - i_{rbss} + i_{rbt}\right) \tag{4.18}$$

The gray areas shown in Figure 4.9b illustrate the contribution of each current of the balancing network to the charge of the flying capacitor. It is clear that the net charge due to i_{rbss} is null, while i_{rbt} introduces a net charge in each switching cycle. The averaged current through the capacitor can be calculated as

$$\tilde{i}_{C2i} = \left(d_1 - d_2\right)i_i + \tilde{i}_{rbt} \tag{4.19}$$

The presence of $\tilde{i}_{rbt}$ allows us to modify the average voltage across C_{2i}, regardless of the duty cycles of the power switches. This gives some independency between the control and modulation of the converter and the recovery from transient states:

$$\frac{d\tilde{v}_{C2i}}{dt} = \frac{1}{C_{2i}}\left(d_1 - d_i\right)i_i + \frac{1}{C_{2i}}\tilde{i}_{rbt} \tag{4.20}$$

Figure 4.8b shows the new averaged circuit in which the balancing network is modeled as a current generator $\tilde{i}_{rbt}$ connected in parallel with the flying capacitor. The amplitude of this current depends on the voltage $\tilde{v}_{C2i}$, and it extinguishes when $\tilde{v}_{C2i}$ reaches the steady-state value of $V_{DC}/2$. The current $\tilde{i}_{rbt}$ is calculated assuming that i_{rbt} is sinusoidal, its amplitude is determined by the merit factor of the tuned network (Q_{rb}), and both gray areas in one commutation cycle do not change. With these considerations, the variations of $\tilde{v}_{C2i}$ can be calculated from (4.20) [12]:

$$\frac{d\tilde{v}_{C2i}}{dt} = 2\frac{C_{rb}}{T_s C_{2i}}\left[1 - e^{(-\alpha d T_s)}\left(\frac{\alpha}{\omega_{rb}}\sin\theta + \cos\theta\right)\right]\left(\frac{V_{DC}}{2} - \tilde{v}_{C2i}\right)$$

$$+2\frac{\omega_{rb}L_{rb}C_{rb}}{T_s C_{2i}}\left(1 - \frac{\alpha}{\omega_{rb}}\right)\left[\frac{\alpha}{\omega_{rb}} - e^{(-\alpha d T_s)}\left(\frac{\alpha}{\omega_{rb}}\cos\theta - \sin\theta\right)\right]\tilde{i}_{Crb0} \qquad (4.21)$$

$$-2\frac{C_{rb}}{T_s C_{2i}}\left[1 - e^{(-\alpha d T_s)}\left(\frac{\alpha}{\omega_{rb}}\sin\theta + \cos\theta\right)\right]\tilde{v}_{Crb0}$$

where $\tilde{i}_{Crb0}$ is the initial value of the current in L_{rb}, $\tilde{v}_{Crb0}$ is the initial value of the voltage across C_{rb}, α is the damping coefficient of the tuned network, ω_{rb} is the damped oscillating frequency, and $\theta = \omega_{rb}.d.T_S$.

The initial values $\tilde{i}_{Crb0}$ and $\tilde{v}_{Crb0}$ vary from one cycle to the next, while $\tilde{v}_{C2i}$ evolves toward the steady state. So it is necessary to calculate how these variables evolve along the transient, in order to complete the circuit model. It is important to take some simplifying hypotheses on the time domain model:

1. The variations of $\tilde{v}_{C2i}$ are much slower than those of $\tilde{v}_{Crb0}$ and $\tilde{i}_{Crb0}$.
2. The balancing network is tuned at the switching frequency,

$$\frac{1}{\sqrt{L_{rb} \cdot C_{rb}}} = \omega_S.$$

3. The merit factor $Q_{rb} = \dfrac{\omega_S L_{rb}}{R_{rb}}$ is much greater than unity.

The following differential equations describe the evolution of the initial values of voltage and current in the balancing network:

$$\frac{d\tilde{i}_{Crb0}}{dt} = -\alpha\frac{\kappa}{\omega_{rb}L_b}\tilde{v}_{C2i} - \alpha\tilde{i}_{Crb0}$$

$$\frac{d\tilde{v}_{Crb0}}{dt} = \alpha\left(\nu + \frac{\alpha}{\omega_{rb}}\kappa\right)\tilde{v}_{C2i} - \alpha\tilde{v}_{Crb0}$$

$\qquad\qquad (4.22)$

where ν and κ are constants that depend on α, ω_{rb}, d, and T_S.

Equations (4.21) and (4.22) represent a third-order system whose roots are

$$q_{1-2} = -\frac{(\alpha + a_1)}{2} \pm \frac{1}{2}\sqrt{(\alpha - a_1)^2 - 4(c_1 a_1 - b_1 a_2)}$$

$$q_3 = -\alpha$$

$\qquad\qquad (4.23)$

with

$$a_1 = 2\frac{C_{rb}}{T_S C_{2i}}\left[1 - e^{(-\alpha d T_s)}\left(\frac{\alpha}{\omega_{rb}}\sin\theta + \cos\theta\right)\right]$$

$$a_2 = 2\frac{\omega_{rb} L_{rb} C_{rb}}{T_S C_{2i}}\left(1 - \frac{\alpha}{\omega_{rb}}\right)\left[\frac{\alpha}{\omega_{rb}} - e^{(-\alpha d T_s)}\left(\frac{\alpha}{\omega_{rb}}\cos\theta - \sin\theta\right)\right]$$

(4.24)

$$b_1 = -\alpha\frac{\kappa}{\omega_{rb} L_b}$$

$$c_1 = \alpha\left(v + \frac{\alpha}{\omega_{rb}}\kappa\right)$$

This system has a fixed root (q_3) that depends only on the damping factor (α) of the balancing network. The other two roots, q_1 and q_2, depend on the network parameters (ω_{rb} and Q_{rb}), the ratio between the capacitance of the balancing network, the ratio C_{rb}/C_{2i}, and the duty cycle of both power switches d. Depending on the selection of the balancing network parameters, the roots q_1 and q_2 may be real or complex conjugates.

4.4.2.1 Root Locus Analysis

The roots of (4.23) fix the speed of convergence of $\tilde{v}_{C2i}$ toward the equilibrium value $V_{DC}/2$. This means that the network parameters determine the dynamic behavior of the voltage $\tilde{v}_{C2i}$. Then, it is very important to analyze the relationship between the parameters and the roots q_1 and q_2.

Equation (4.24) shows that the ratio C_{rb}/C_{2i} directly affects the values of q_1 and q_2. The root locus for varying C_{rb}/C_{2i} is presented in Figure 4.10 for a given merit factor and duty cycle. The locus also shows that q_3 maintains a value higher than q_1 and q_2. Moreover, for low values of C_{rb}/C_{2i}, $(\alpha - a_1)^2 \gg 4(c_1 a_1 - b_1 a_2)$; all the roots are real and they can be simplified. In this situation $q_1 \cong -a_1$, $q_2 \cong -\alpha = q_3$, and q_1 is the dominant pole that guides the evolution of $\tilde{v}_{C2i}$ as an exponential with a time constant equal to $1/q_1$. For higher values of C_{rb}/C_{2i}, q_1 and q_2 become complex. In this case the speed of variation of $\tilde{v}_{C2i}$ is comparable with the changes in $\tilde{v}_{Crb0}$ and $\tilde{i}_{Crb0}$, and hypothesis 1 is no longer satisfied. Then, it is important to design the balancing network in a way to have real poles in order to determine precisely the evolution of $\tilde{v}_{C2i}$.

There exist several trade-offs in the selection of the balancing network parameters. Figure 4.11 shows the real and imaginary parts of the roots q_1 and q_2 versus the ratio C_{rb}/C_{2i}, for different values of the merit factor of the network and a fixed duty cycle. For low values of the merit factor, the

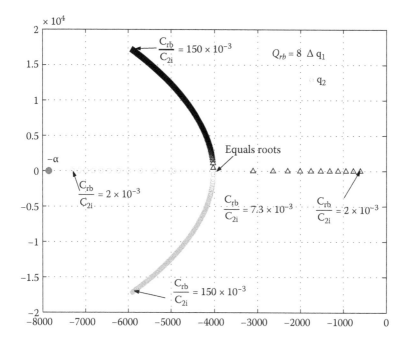

Figure 4.10 Root locus of q_1 and q_2 as function of C_{rb}/C_{2i}.

range of C_{rb}/C_{2i} for which the roots are real is wider. But at the same time, the value of q_1 is low, and so the time of convergence of $\tilde{v}_{C2i}$ to its steady state is high. Then, it is necessary to increase C_{rb}/C_{2i} to speed up the convergence. It is possible to choose a design criterion for a fast and predictable convergence of $\tilde{v}_{C2i}$ from the graphs in Figure 4.11. The roots q_1 and q_2 should be real and similar. Then it is advisable to choose a high value for Q_{rb} (this also implies less losses in the balancing network) and adopt a ratio C_{rb}/C_{2i} small enough to guarantee that the roots are real.

Figure 4.12 shows the start-up charge of the flying capacitor for the actual circuit and that predicted by the model given by (4.21) and (4.22). Four cases are shown for different balancing networks. They all have the same resonance frequency (ω_s) and the same ratio C_{rb}/C_{2i}. They differ in the merit factor of the balancing network.

In the precedent analysis the duty cycle was assumed to be constant. But in a modulated converter it will vary in a range between 0 and 0.5. It is not easy to find a root locus for variable duty cycle, but it is possible to determine for which value of C_{rb}/C_{2i} the roots are real and coincident for a varying duty cycle. These curves are presented in Figure 4.13; they show the limit between real and complex roots. The points below these curves guarantee real roots in all duty cycle ranges.

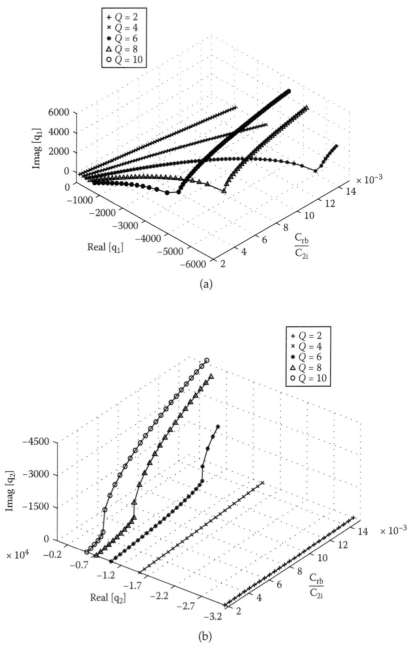

Figure 4.11 Real and imaginary part of q_1 (a) and q_2 (b) versus C_{rb}/C_{2i} (C_{2i} = 160 μF, d = 0.2).

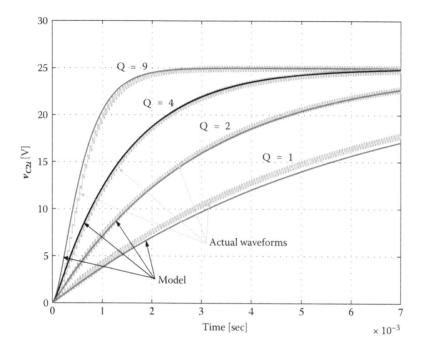

Figure 4.12 Examples of actual waveforms and model of the $\tilde{v}_{C2i}$ at start-up ($C_{rb} =$ 795 nF, $L_{rb} = 77$ µH, $C_{2i} = 160$ µF, $V_{DC} = 50$ V).

4.5 Summary

The flying capacitor converter is a symmetric topology with common DC bus in which it is easy to increase the number of voltage levels. In this chapter it was shown that a phase-shifted carrier modulation uses all the redundant states and helps to maintain the voltage balance on the flying capacitors in steady state. However, this modulation scheme cannot address the initial charging condition at start-up or the transient voltage deviations on the flying capacitors from their steady-state values. For this, the injection of balancing currents using a tuned balancing network was proposed.

A time domain analysis was performed to see the effects of connecting the tuned network. This analysis allows us to describe the evolution of the flying capacitor voltages toward their steady state. The model can be approximated to a first-order system or an overdamped second-order system depending on the chosen values of the capacitance ratio C_{rb}/C_{2i} and the merit factor Q_{rb}. Increasing Q_{rb} implies the increment of the convergence speed, but at the same time it requires decreasing the ratio C_{rb}/C_{2i} in order to satisfy the simplifying hypothesis.

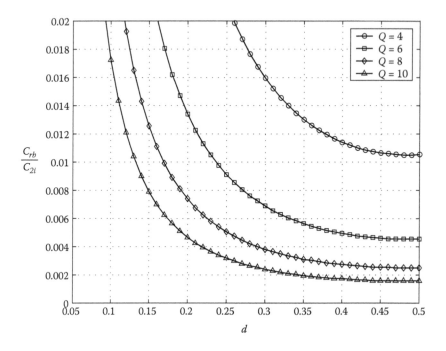

Figure 4.13 Border lines between real and complex roots for varying duty cycles.

References

1. T.A. Meynard, H. Foch. Multi-Level Conversion: High Voltage Choppers and Voltage-Source Inverters. In *IEEE Power Electronics Specialists Conference (PESC'92)*, Toledo, Spain, June 29–July 3, 1992, vol. 1, pp. 397–403.

2. D.G. Holmes, T.A. Lipo. *Pulse Width Modulation for Power Converters: Principles and Practice.* IEEE Press, Piscataway, NJ, 2003.

3. C. Feng, J. Liang, V.G. Agelidis. A Novel Voltage Balancing Control Method for Flying Capacitor Multilevel Converters. In *IEEE Annual Conference of the Industrial Electronics Society (IECON'03)*, Roanoke, VA, November 2–6, 2003, pp. 1179–1184.

4. M.A. Severo Mendes, Z.M. Assis Peixoto, P.F. Seixas, P. Donoso-Garcia. A Space Vector PWM Method for Three-Level Flying-Capacitor Inverters. In *Power Electronics Specialists Conference (PESC'01)*, Vancouver, Canada, June 17–21, 2001, pp. 182–187.

5. F. Defaÿ, A.M. Llor, M. Fadel. A Predictive Control with Flying Capacitor Balancing of a Multicell Active Power Filter. *IEEE Transactions on Industrial Electronics*, 55(9), 3212–3220, 2008.

6. L. Xu, V.G. Agelidis. A VSC Transmission System Using Flying Capacitor Multilevel Converters and Selective Harmonic Elimination PWM Control. *International Journal of Emerging Electric Power Systems*, 5(2), 1–16, 2006.

7. W.K. Lee, S.Y Kim, J.S. Yoon, D.H. Baek. A Comparison of the Carrier-Based PWM Techniques for Voltage Balance of Flying Capacitor in the Flying Capacitor Multilevel Inverter. In *Applied Power Electronics Conference and Exposition (APEC'06)*, March 19–23, 2006, pp. 1653–1658.
8. T.A. Meynard, M. Fadel, N. Aouda. Modeling of Multilevel Converts. *IEEE Transactions on Industrial Electronics*, 44(3), 356–364, 1997.
9. X. Yuan, H. Stemmler, I. Barbi. Self-Balancing of the Clamping-Capacitor-Voltages in the Multilevel Capacitor-Clamping-Inverter under Sub-Harmonic PWM Modulation. *IEEE Transactions on Power Electronics*, 16(2), 256–263, 2001.
10. T.A. Meynard, H. Foch, P. Thomas, J. Courault, R. Jakob, M. Nahrstaedt. Multicell Converters: Basic Concepts and Industry Applications. *IEEE Transactions on Industrial Electronics*, 49(5), 955–964, 2002.
11. B.P. McGrath, D.G. Holmes. Analytical Modeling of Voltage Balance Dynamics for a Flying Capacitor Multilevel Converter. *IEEE Transactions on Power Electronics*, 23(2), 543–550, 2008.
12. S.A. González, M.I. Valla, C.F. Christiansen. Design of a Tuned Balancing Network for Flying Capacitor Multilevel Converters. In *IEEE Power Electronics Specialists Conference (PESC'05)*, Recife, Brazil, June 1–7, 2005, pp. 1046–1051.
13. R. Erickson, D. Macsimovik. *Fundamentals of Power Electronics*, 2nd ed. Kluwer Academic, Norwell, MA, 2001.

chapter 5

Cascade Asymmetric Multilevel Converter

5.1 Introduction

Chapters 2 through 4 have shown that the diode-clamped and flying capacitor multilevel converters have some limitations to increase the number of voltage levels. The number of components increases exponentially with the number of levels. Thus, the complexity, volume, stored energy, and consequently, their cost greatly increase [1]. So, a new trend in the development of multilevel converters is to improve the ratio between the number of voltage levels at the output and the number of states of the converter. The asymmetric topologies generally satisfy this objective. The well-known cascaded cell multilevel converter satisfies this goal, but at the expense of using isolated power sources for each stage. There exist some alternatives to combine different topologies maintaining a common DC bus. The connection of two piled-up flying capacitor cells connected in cascade with a neutral point clamped (NPC) converter cell was discussed in Chapter 2. But this topology does not obtain a significant reduction on the component count. Following this idea a new five-level topology with a common DC bus appeared a few years ago. It was named cascade asymmetric multilevel converter (CAMC) [2].

The CAMC has two asymmetric cascaded stages. The high-voltage (HV) stage is followed by the low-voltage (LV) stage. It has characteristics similar to those of the two-stage hybrid multilevel converter (HMC) with double-voltage progression, but it offers the advantage of a single-voltage source. The CAMC has a reduced number of capacitors; it has less redundant states, so it improves the ratio between voltage levels and possible states, when compared with the classical symmetric topologies analyzed in Chapters 3 and 4.

A hybrid modulation strategy for the CAMC is presented in this chapter. It allows traveling along all the states of the converter maintaining the voltage balance on all the capacitors. A further analysis of the low-frequency spectrum of the DC bus current shows the charge balance at steady state for different load conditions. In this analysis, two components of the DC bus current are identified: a common mode current and a differential mode current. The differential mode current defines the design of the DC bus capacitors and the related energy storage. The chapter ends with a short comparative analysis of the different topologies that have been presented in the book.

5.2   General Characteristics of the CAMC

Figure 5.1 shows one leg of the CAMC. Here both stages in cascade can be seen. The HV stage is formed with two stacked basic cells that are fed by the DC voltage. The LV stage is a three-level flying capacitor topology connected between nodes w and z, and it is fed with half of the DC voltage.

The capacitors C_1 and C_2 have the same value, and they fix the blocking voltage, on the switches S_{1w}, S_{1z}, and their complements, to $V_{DC}/2$. The HV stage commutates at low frequency so its switches may be implemented with integrated gate-commutated thyristor (IGCT) devices in order to withstand higher-voltage levels. It was described in Chapter 2 that the switching strategy in the HV stage should impose a voltage equal to $V_{DC}/2$ between the nodes w and z. This is the voltage source for the LV stage, so the blocking voltage for the LV stage switches should be $V_{DC}/4$. This stage is controlled with a pulse width modulation switching at high frequency, so its switches are implemented with insulated gate bipolar transistor (IGBT) devices. This converter requires only three switching functions to achieve the five voltage levels. The HV stage has only one switching function (s_1) to control switches S_{1w} and S_{1z}. The LV stage is a three-level flying capacitor stage, so S_2 and S_3 are controlled by the switching functions s_2 and s_3, to build up and maintain the voltage amplitude

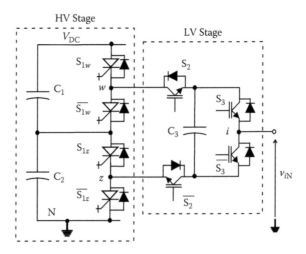

Figure 5.1 One leg of the CAMC.

equal to $V_{DC}/4$ on the flying capacitor (C_3) and to obtain the desired voltage at the output. Then, the leg voltage v_{iN} is obtained summing up the contribution of both stages.

5.2.1 Modulation Strategy

The CAMC is modulated with a hybrid strategy. It combines the low frequency of the HV stage with the PWM for the LV stage. The HV stage generates a square wave with an amplitude $V_{DC}/2$ on the leg voltage v_{iN}. The LV stage is modulated in such a way to obtain a sinusoid on the leg voltage v_{iN} after summing the contribution from the HV stage.

Figure 5.2 shows the voltage waveforms generated by each stage with the proposed modulation strategy. Since the LV stage is a three-level flying capacitor converter, it requires a phase-shifted carrier pulse width modulation (PSPWM) in order to preserve the charge balance of the flying capacitor. This modulation is shown in Figure 5.2a. It requires two carriers with an amplitude A_p and a 180° phase shift between them. The modulating signal for the LV stage is obtained subtracting a square wave of amplitude A_p from the desired sinusoid, and it is also shown in Figure 5.2a. It may be expressed as

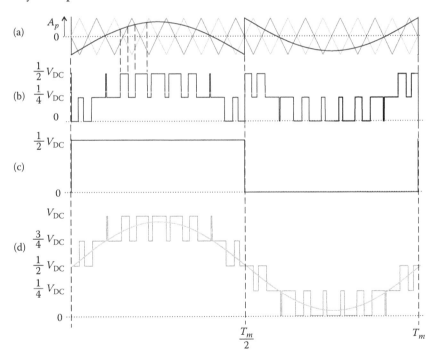

Figure 5.2 Modulation strategy.

$$v_{mi} = 2mA_p \sin\left(\omega_m t - \psi_i\right) - A_p\left(2s_{1i} - 1\right) \tag{5.1}$$

where m is the modulation index that is always less than one, ω_m is the frequency of the modulating signal, and ψ_i is the initial phase of the voltage corresponding to leg i.

The modulating signal given in (5.1) is compared with both carriers generating the driving signals for S_2 and S_3. The result is the three-level voltage waveform shown in Figure 5.2b. The HV stage is directly driven by the zero crossing of the modulating signal at a frequency ω_m, leading to the waveform shown in Figure 5.2c. The leg voltage is obtained as the sum of the voltages generated by each stage. As was shown in Chapter 2, it results in

$$v_{iN} = \frac{V_{DC}}{2} s_1 + \frac{V_{DC}}{4}\left(s_2 + s_3\right) \tag{5.2}$$

Finally, Figure 5.2d shows the resultant five-level leg voltage v_{iN}.

5.2.2   Averaged Voltage

If the carrier period (T_S) is sufficiently small, it can be assumed that the modulating signal is constant along T_S. Also, if it is considered that the modulation index is less than 1, then the fundamental component of v_{iN} may be calculated taking its average in each switching cycle T_S. The sinusoidal trace in Figure 5.2d shows the averaged value of the leg voltage.

Taking the average of (5.2) over T_S and recalling that the first term is constant, since it is a square wave at the modulating frequency, the averaged value of the leg voltage can be expressed as

$$\tilde{v}_{iN} = \frac{V_{CC}}{2} s_{1i} + \frac{V_{CC}}{4}\left(\frac{1}{T_S}\int_{T_S}\left(s_{2i} + s_{3i}\right)d\tau\right) \tag{5.3}$$

Then the averaged value $\tilde{v}_{iN}$ is obtained from the average of the switching functions s_{2i} and s_{3i}. Figure 5.3 shows a detail of two carrier cycles and the modulating signal for one leg of the converter, together with the resulting switching functions s_{2i} and s_{3i}.

The average of the switching functions defines the duty cycle of the switches:

$$d_i = \frac{1}{T_S}\int_{T_S} s_i d\tau \tag{5.4}$$

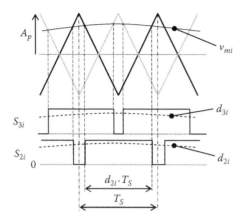

Figure 5.3 PSPWM modulation ($n = 3$). Switching functions s_2 and s_3.

Then, the averaged voltage at the leg output equals

$$\tilde{v}_{iN} = \frac{V_{CC}}{2}s_{1i} + \frac{V_{CC}}{4}(d_{2i} + d_{3i}) \tag{5.5}$$

Looking to the switching functions in Figure 5.3, it is easy to determine that the time intervals $d_{2i}.T_S$ and $d_{3i}.T_S$ equal the time in which the modulating signal is higher than the corresponding carrier. Assuming that v_{mi} is constant along one switching cycle, the duty cycles may be approximated by

$$d_{2i}(t) = d_{3i}(t) \cong \frac{1}{2} + \frac{1}{2}\frac{v_{mi}(t)}{Ap} \tag{5.6}$$

Replacing v_{mi} by the value given in (5.1), the duty cycles are

$$d_{2i} = d_{3i} = \frac{1}{2} + \frac{1}{2}\frac{v_{mi}}{Ap} = 1 + m.\sin(\omega_m t - \psi_i) - s_{1i} \tag{5.7}$$

Finally, replacing (5.7) in (5.5), the averaged leg voltage is obtained:

$$\tilde{v}_{iN} = \frac{V_{CC}}{2} + m\frac{V_{CC}}{2}\sin(\omega_m t - \psi_i) \tag{5.8}$$

Equation (5.8) shows that the fundamental component has a linear dependence with the modulation index.

5.3 CAMC Three-Phase Inverter

Up to now we have described one leg of the CAMC converter, but this is immediately extended to a three-phase converter connecting three legs to a common DC bus. In this way we obtain a multilevel converter that is attractive for industry applications. Voltage source multilevel converters are available for different industrial applications, such as electric drives in medium voltage and high power [3,4]. Most of the converters found commercially are three-level converters like the NPC converter (three-level DCMC) [5,6]. Four-level flying capacitor multilevel converters can also be found in electric drives [6]. Among them the CAMC topology offers five voltage levels with low complexity and a simple modulation scheme. So it becomes attractive in several applications of a three-phase inverter, electric drives, or power conditioning in distribution lines.

Figure 5.4 shows the implementation of a three-phase CAMC in a general three-wire application. The DC voltage source V_{DC} represents an energy storage system like a battery bank [7,8], or may correspond to a back-to-back connection with another converter. The load may represent an AC motor in a drive application, or the electric network when working as a distribution static compensator (DSTATCOM), active power filter, or dynamic voltage restorer (DVR).

The modulation scheme presented in the previous section can be applied to the converter shown in Figure 5.4. In this case, there are three modulating signals that correspond to the three phases. The modulating signal for the LV stage of each phase has the form given in (5.1) (with $i = a, b, c$), and the initial phases for each signal are $\psi_a = 0$, $\psi_b = 2\pi/3$, and $\psi_c = -2\pi/3$, respectively. This generates a balanced three-phase voltage system, as shown in Figure 5.5, in which the DC bus voltage is 25 kV and the converter operates with a modulation index $m = 0.9$ and drives a balanced load. The upper figure shows the leg voltages (v_{iN}), while the phase voltages (v_{io}) are shown in the lower figure.

It is important to analyze the currents in the DC bus as a function of the load, especially in those applications where the DC voltage depends on capacitors C_1 and C_2, like in a back-to-back connection or in conditioning devices that deal only with reactive power and have no energy source on the DC bus.

5.3.1 Averaged Currents in the DC Bus

5.3.1.1 i_{DC} Current

The three HV stages of the CAMC shown in Figure 5.4 are connected to the same DC bus. Then, the three phases of the converter contribute to the current through C_1 and C_2. Looking at Figure 5.4 it is easy to determine that the current in the DC bus (i_{DC}) is the sum of the currents through the

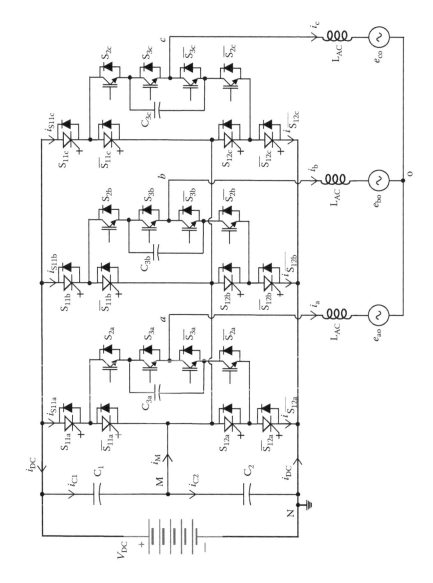

Figure 5.4 CAMC converter in a general application.

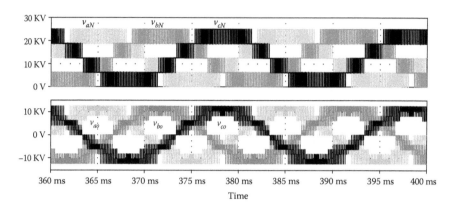

Figure 5.5 Leg and phase voltages of the CAMC.

upper switches of each leg (i_{S11i}), or the sum of the currents through the lower switches of each leg ($i_{\overline{S12i}}$), with $i = a$, b, and c:

$$i_{DC} = -\left(i_{S11a} + i_{S11b} + i_{S11c}\right) = -\left(i_{\overline{S12a}} + i_{\overline{S12b}} + i_{\overline{S12c}}\right) \tag{5.9}$$

In steady state, i_{DC} is a periodic waveform whose harmonic spectrum contains high-frequency components generated by the PSPWM of the LV stage, and low-order harmonics that result from the interaction of the square wave of the HV stage and the harmonics of the load. The high-order harmonics have a small impact on the average voltage on the capacitors C_1 and C_2. The voltage ripple is mainly determined by the low-order harmonics. These components can be calculated using the average current over one switching period T_S, similarly as was done with the average voltage at the AC side.

It is necessary to find the relationship between the currents through the HV stage switches and the load current ($i_{(i)}$), averaging the switching action of the LV stage. The current through any switch of the LV stage can be written as the product of the load current $i_{(i)}$ and the corresponding switching function, the same as was analyzed in Chapter 4. The current through any switch of the LV stage S_{ji} (with $j = 2$, 3 and $i = a$, b, c) is written as

$$i_{Sji} = s_{ji}i_{(i)} \tag{5.10}$$

The current through the upper switch S_{11i} can also be expressed as the product of the current through S_{2i} of the LV stage and its own switching function s_{1i}. Using the expression in (5.10), the current through S_{11i} can be written as

$$i_{S11i} = s_{1i}\left(s_{2i}i_{(i)}\right) \tag{5.11}$$

The current given in (5.11) has high-order harmonics given mainly by s_{2i} and low-order harmonics that result from the interaction of the low-frequency commutation s_{1i} and the load current.

The average value of i_{S11i} over T_S is obtained averaging the current through the LV stage switch (S_{2i}), since s_{1i} is constant along T_S:

$$\overline{i_{S11i}} = s_{1i}\left(\frac{1}{T_S}\int_{T_S} s_{2i}i_{(i)}d\tau\right) \tag{5.12}$$

Assuming also that the load current is almost constant along the averaging interval T_S, the averaged current through S_{2i} is the product between the load current $i_{(i)}$ and the averaged switching function s_{2i}, which is the duty cycle of the switch. For a sinusoidal modulation, recalling (5.7), the averaged current through S_{2i} has the expression

$$\overline{i_{S2i}}(t) = i_{(i)}\left[1 + m\cdot\sin\left(\omega_m t - \psi_i\right) - s_{1i}\right] \tag{5.13}$$

where $i = a, b$, and c and the initial phases are $\psi_a = 0$, $\psi_b = 2\pi/3$, and $\psi_c = -2\pi/3$.

The averaged current through the upper switch can be calculated introducing (5.13) in (5.12). Since the product of a switching function by itself results in the same switching function, the current is expressed as

$$\overline{i_{S11i}} = s_{1i}\cdot i_{(i)}m.\sin\left(\omega_m t - \psi_i\right) \tag{5.14}$$

To obtain the spectrum of this current, it is necessary to have the Fourier series of the switching function and the load current. The Fourier series of the switching function s_{1i} of each leg of the converter is

$$s_{1i} = \frac{1}{2} + \frac{2}{\pi}\sum_{j(odd)}\frac{1}{j}\sin\left[j\cdot(\omega_m t - \psi_i)\right] \tag{5.15}$$

Introducing this expression in (5.14), the averaged current in each upper switch of the HV stage results in

$$\overline{i_{S11i}} = \frac{m\cdot i_{(i)}}{2}\sin\left(\omega_m t - \psi_i\right) + \frac{m\cdot i_{(i)}}{\pi}\sum_{j=1}\frac{1}{j}\cdot\begin{bmatrix}\cos\left[(j-1)\cdot(\omega_m t - \psi_i)\right] \\ -\cos\left[(j+1)\cdot(\omega_m t - \psi_i)\right]\end{bmatrix} \tag{5.16}$$

Assuming that the load current is balanced with odd harmonics different than triple, the Fourier series of each load phase current is

$$i_{(i)} = \sum_{h(odd)} I_h \sin\left[h \cdot (\omega_m t - \psi_i) - \varphi_h\right] \tag{5.17}$$

where $I_h = I_1/h$ and I_1 is the amplitude of the fundamental component of the load current.

Finally, introducing (5.17) in (5.16) and summing the currents of the three upper switches, the averaged current in the DC bus results in

$$\bar{i}_{DC} = -\frac{1}{4} m \sum_{h=1} I_h \begin{bmatrix} \cos\left((h-1) \cdot \omega_m t - \varphi_h\right) \\ +\cos\left((h-1) \cdot (\omega_m t - \tfrac{2}{3}\pi) - \varphi_h\right) \\ +\cos\left((h-1) \cdot (\omega_m t + \tfrac{2}{3}\pi) - \varphi_h\right) \\ -\cos\left((h+1) \cdot \omega_m t - \varphi_h\right) - \\ -\cos\left((h+1) \cdot (\omega_m t - \tfrac{2}{3}\pi) - \varphi_h\right) \\ -\cos\left((h+1) \cdot (\omega_m t + \tfrac{2}{3}\pi) - \varphi_h\right) \end{bmatrix}$$

$$-\frac{m}{2\pi} \sum_{h=1} I_h \sum_{j=1} \frac{1}{j} \left\{ \begin{bmatrix} \sin\left[(h-j+1) \cdot \omega_m t - \varphi_h\right] \\ +\sin\left[(h+j-1) \cdot \omega_m t - \varphi_h\right] \\ +\sin\left[(h-j+1) \cdot (\omega_m t - \tfrac{2}{3}\pi) - \varphi_h\right] \\ +\sin\left[(h+j-1) \cdot (\omega_m t - \tfrac{2}{3}\pi) - \varphi_h\right] \\ +\sin\left[(h-j+1) \cdot (\omega_m t + \tfrac{2}{3}\pi) - \varphi_h\right] \\ +\sin\left[(h+j-1) \cdot (\omega_m t + \tfrac{2}{3}\pi) - \varphi_h\right] \end{bmatrix} \\ -\begin{bmatrix} \sin\left[(h-j-1) \cdot \omega_m t - \varphi_h\right] \\ +\sin\left[(j+h+1) \cdot \omega_m t - \varphi_h\right] \\ +\sin\left[(h-j-1) \cdot (\omega_m t - \tfrac{2}{3}\pi) - \varphi_h\right] \\ +\sin\left[(h+j+1) \cdot (\omega_m t - \tfrac{2}{3}\pi) - \varphi_h\right] \\ +\sin\left[(h-j-1) \cdot (\omega_m t + \tfrac{2}{3}\pi) - \varphi_h\right] \\ +\sin\left[(j+h+1) \cdot (\omega_m t + \tfrac{2}{3}\pi) - \varphi_h\right] \end{bmatrix} \right\} \tag{5.18}$$

where m is the modulation index of the LV stage, I_h is the amplitude of the hth harmonics of the load current (h is odd and a nonmultiple of 3), φ_h is the phase displacement of the load current, and j represents odd harmonics.

5.3.1.2 Common Mode and Differential Mode Currents

The current in the DC bus, which was calculated in the previous section, has a complex expression given in (5.18). In order to have a better insight of this current, it is separated in two components with different characteristics. They are a common mode current ($\bar{i}_C$) and a differential mode current ($\bar{i}_D$). To identify these components, it is necessary to analyze the current in the midpoint between both capacitors C_1 and C_2, identified as node M (i_M) in Figure 5.4. This node receives currents from the three legs of the converter and equals the current difference between both capacitors, so

$$i_{C1} - i_{C2} = i_M = i_{Ma} + i_{Mb} + i_{Mc} \tag{5.19}$$

The averaged value of i_M equals the sum of the average currents of each leg of the converter:

$$\bar{i}_M = \bar{i}_{Ma} + \bar{i}_{Mb} + \bar{i}_{Mc} \tag{5.20}$$

At the same time, each leg current i_{Mi} can be written as a function of the current in the individual switches:

$$\bar{i}_{Mi} = \bar{i}_{S12i} - \bar{i}_{S11i} = -\bar{i}_{S2}s_{1i} + \bar{i}_{S2i}(1 - s_{1i}) \tag{5.21}$$

Introducing (5.21), for each phase in (5.20) and operating in a similar way as was done in the previous section, the current in node M is obtained:

$$\bar{i}_M = \frac{m}{\pi}\sum_{h=1}^{}I_h\sum_{j=1}^{}\frac{1}{j}\left\{ \begin{bmatrix} \sin\left[(h-j+1)\cdot(\omega_m t)-\varphi_h\right] \\ +\sin\left[(h+j-1)\cdot(\omega_m t)-\varphi_h\right] \\ +\sin\left[(h-j+1)\cdot(\omega_m t-\frac{2}{3}\pi)-\varphi_h\right] \\ +\sin\left[(h+j-1)\cdot(\omega_m t-\frac{2}{3}\pi)-\varphi_h\right] \\ +\sin\left[(h-j+1)\cdot(\omega_m t+\frac{2}{3}\pi)-\varphi_h\right] \\ +\sin\left[(h+j-1)\cdot(\omega_m t+\frac{2}{3}\pi)-\varphi_h\right] \end{bmatrix} - \begin{bmatrix} \sin\left[(h-j-1)\cdot(\omega_m t)-\varphi_h\right] \\ +\sin\left[(h+j+1)\cdot(\omega_m t)-\varphi_h\right] \\ +\sin\left[(h-j-1)\cdot(\omega_m t-\frac{2}{3}\pi)-\varphi_h\right] \\ +\sin\left[(h+j+1)\cdot(\omega_m t-\frac{2}{3}\pi)-\varphi_h\right] \\ +\sin\left[(h-j-1)\cdot(\omega_m t+\frac{2}{3}\pi)-\varphi_h\right] \\ +\sin\left[(h+j+1)\cdot(\omega_m t+\frac{2}{3}\pi)-\varphi_h\right] \end{bmatrix} \right\} \quad (5.22)$$

This averaged current equals the double of the second term in (5.18) [9]. Then, the differential current is given by the second term in (5.18), and the common mode current is given by the first term in (5.18). Then,

$$\bar{i}_{DC} = \bar{i}_C + \frac{\bar{i}_M}{2} \quad (5.23)$$

Figure 5.6 presents a steady-state model of the DC side of the converter, identifying both components of the current. It clearly shows that the common mode current $\bar{i}_C$ circulates through the voltage source, which feeds the DC bus, while the differential current circulates through capacitors C_1 and C_2. In those applications that do not require active power, like reactive power or harmonics compensation, there is no voltage source and the common mode current also circulates through the capacitors.

5.3.2   Common Mode Current

The first term of (5.18) is the common mode current $\bar{i}_C$. It is composed of a DC component (I_{DC}), which corresponds to the active power demanded by

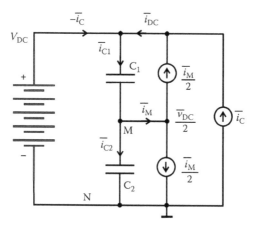

Figure 5.6 Steady-state model for currents in the DC bus.

the load plus triple harmonics defined by the harmonics contents of the load. When the load current $i_{(i)}$ defined in (5.17) corresponds to a balanced and linear load, only the fundamental component ($h = 1$) will be present with a phase angle φ_1. Then, the DC component of the common mode current is

$$\bar{i}_C = -\frac{3}{4} m \cdot I_1 \cos(\varphi_1) \tag{5.24}$$

This current corresponds to the active power consumed by the load in the AC side of the converter. When the load is balanced but nonlinear, like a six-pulse rectifier, the load current has odd harmonics, like $h = 5, 7, 11, 13, 17, 19$, and so forth. Now the common mode current will have the DC value plus even triple harmonics. For example, the load harmonics $h = 5$ and $h = 7$ generate sixth harmonics in the common mode current with the following expression:

$$\bar{i}_{C6} = \frac{3}{4} m \cdot I_6 \cos(6\omega_m t - \varphi_6) \tag{5.25}$$

The amplitude I_6 and the angle φ_6 depend on the load harmonics that generate them:

$$I_6 = \sqrt{\left(I_5 \cos\varphi_5 - I_7 \cos\varphi_7\right)^2 + \left(I_5 \sin\varphi_5 - I_7 \sin\varphi_7\right)^2} \tag{5.26}$$

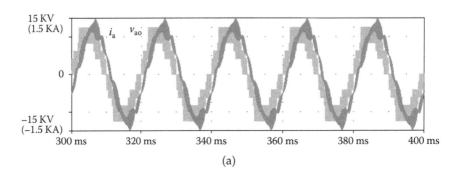

(a)

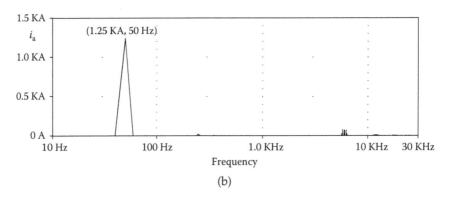

(b)

Figure 5.7 CAMC as inverter (load cos φ = 0.9): (a) phase voltage (2 KV/div) and current (0.2 KA/div), (b) load current spectrum, (c) current in the DC bus, and (d) current in the DC bus spectrum.

and

$$\varphi_6 = -\tan^{-1}\left(\frac{I_5 \cos\varphi_5 - I_7 \cos\varphi_7}{I_5 \sin\varphi_5 - I_7 \sin\varphi_7}\right) \qquad (5.27)$$

In a similar way $h = 11$ and $h = 13$ generate common mode 12th harmonics, and so on. The load harmonics also generate other even harmonics, but all that are not triple are canceled due to the balanced nature of the load.

Let us see a first example, where the CAMC is feeding a linear inductive load of 20 MVA with cos φ = 0.9. The DC bus is fed by a 25 kV voltage source. The modulation index is set to $m = 0.9$ generating a fundamental phase voltage equal to 8.1 kV$_{RMS}$. Figure 5.7a shows the phase voltage (v_{ao})

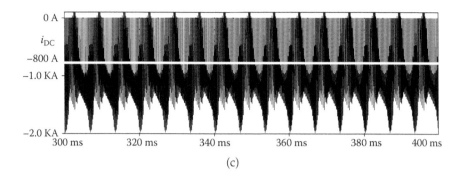

(c)

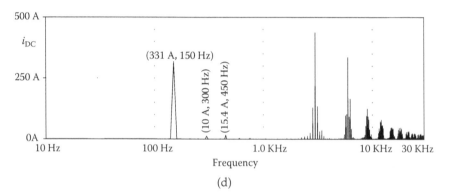

Frequency

(d)

Figure 5.7 (Continued)

together with its corresponding current (i_a), which is slightly delayed. The spectrum of this current is shown in Figure 5.7b, indicating a fundamental component of 1.2 kA at 50 Hz and high-order harmonics around the double of the carrier frequency (3 kHz). Figure 5.7c presents the current in the DC bus i_{DC} with a DC value of 800 A. Figure 5.7d shows the harmonic spectrum of i_{DC}. The common mode current equals the mean value and low order harmonics are odd and triple as predicted in the previous analysis. Also, the high-frequency components due to the PWM can be observed.

A second example employs the same converter feeding a six-pulse diode rectifier. The results of this example are presented in Figure 5.8. Figure 5.8a shows the line current of the CAMC together with its spectrum. The spectrum shows the presence of 5th, 7th, 11th, and 13th low-order harmonics. Figure 5.8b shows the current on the DC side of the CAMC together with its spectrum. The DC value of 860 A corresponds to the active power demanded by the load. In this case there are common mode low-frequency harmonics of the 6th, 12th, and 18th orders.

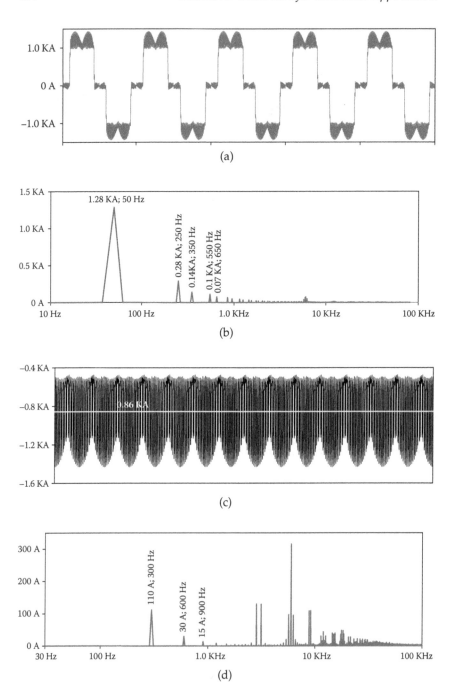

Figure 5.8 (a) Phase current of a six-pulse rectifier, (b) its spectrum. (c) DC common mode current and (d) its spectrum.

5.3.3 Differential Mode Harmonic Currents

As it was previously analyzed, the differential component of the DC bus current is determined by the second term of (5.18). These harmonics are the result of the interaction between the load current harmonics and the harmonics of the switching function (s_{1i}) of the HV stage. All these components may be associated with four groups, λ_1, λ_2, λ_3, and λ_4. Each of them determines a three-phase set at different harmonics. They are defined by

$$\lambda_1 = (h-j+1), \quad \lambda_2 = (h+j-1),$$
$$\lambda_3 = (h-j-1), \quad \lambda_4 = (h+j+1)$$

(5.28)

where j is odd and corresponds to s_{1i}, and h corresponds to load current harmonics. The only harmonic components that may be present in the differential current are those combinations of h and j that generate zero-sequence harmonics, that is, a multiple of three,

$$\lambda_{1,2,3,4} = \pm 3(2k-1)$$

(5.29)

with k a positive integer. The other combinations of h and j generate positive or negative sequences that cancel when they are summed. Taking all this into account, the differential current may be written as

$$\frac{\overline{i_M}}{2} = \frac{m}{2\pi} \sum_{h=1} I_h \sum_{k=1} b_h^{(3(2k-1))} \sin\left[3(2k-1)\cdot\omega_m t - \beta_h^{(3(2k-1))} \right]$$

(5.30)

The amplitude of the load current harmonics has the following relationship: $I_h = I_1/h$; $b_h^{(3(2k-1))}$ represents the amplitude of each differential harmonic, and $\beta_h^{(3(2k-1))}$ is the relative phase of each differential harmonic. The analytical expression of these coefficients is quite complex to arrive to some conclusion; thus, in some cases it is better to perform a graphical analysis.

Figure 5.9 shows the coefficients $b_h^{(3(2k-1))}$ as a function of the phase displacement of the load (φ). The figure shows the coefficients related to the main harmonic components $h = 1, 5, 7,$ and 11. The correspondent coefficients are $b_1^{(3(2k-1))}$, $b_5^{(3(2k-1))}$, $b_7^{(3(2k-1))}$, and $b_{11}^{(3(2k-1))}$, respectively. These coefficients show the different harmonics (k), which are present in the differential current due to each harmonic in the load current. Figure 5.9 shows the first six components.

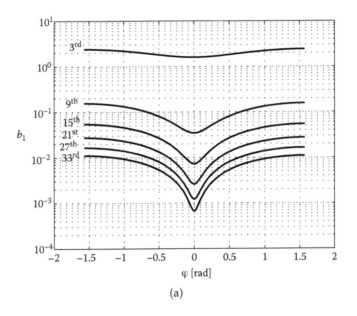

(a)

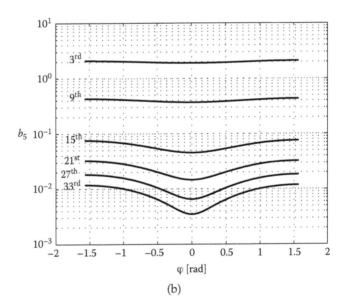

(b)

Figure 5.9 Coefficients $b_1^{(3(2k-1))}$ for (a) $h = 1$, (b) $h = 5$, (c) $h = 7$, and (d) $h = 11$.

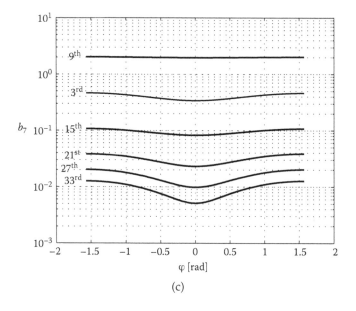

(c)

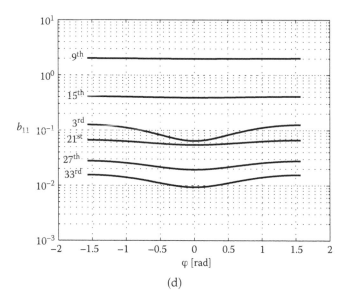

(d)

Figure 5.9 (Continued)

Generally, all the coefficients have greater amplitude when the load is purely reactive, $\varphi = \pm \pi/2$ rad. This is particularly important for the coefficients $b_1^{(3(2k-1))}$ corresponding to the fundamental component of the load current. Figure 5.9a clearly shows that the amplitude of the third harmonics is 15.4 times higher than that of the ninth. On the contrary, for a pure resistive load, $\varphi = 0$ rad, the third harmonics is 46.2 times higher than that of the ninth. Figure 5.9a also shows that this difference increases when higher harmonics are taken into account.

Along changes in the load characteristics, there are no significant modifications between the coefficients values when the current harmonics of the load are considered. The coefficients of the fifth harmonics, $b_5^{(3(2k-1))}$, are shown in Figure 5.9b. In this case the third harmonics is only 4.8 times higher than the ninth for $\varphi = \pm \pi/2$ rad, and 5.16 times higher than the ninth for $\varphi = 0$ rad. Similar considerations may be obtained for the coefficients $b_7^{(3(2k-1))}$ shown in Figure 5.9c. In this case, the highest amplitude corresponds to the ninth harmonics. In the case of $b_{11}^{(3(2k-1))}$ shown in Figure 5.9d, the 15th harmonics presents the highest amplitude. In both cases this is the result of the interaction of the harmonic currents of the load with the fundamental component of the generated voltage.

It should be pointed out that even when the highest values of the coefficients $b_h^{(3(2k-1))}$ have similar magnitudes, their actual weight on the DC current is affected by the load harmonics amplitude, which is attenuated by a factor $1/h$. Analyzing the phase coefficients $\beta_h^{(3(2k-1))}$, they are equal to zero for a balanced load [9].

Returning to the examples studied in the previous section, the differential component of the current is now analyzed with the current through one capacitor, for example, C_1. Figure 5.10a shows the current through C_1, while Figure 5.10b shows its spectrum. This current presents harmonics of third and ninth orders, and other odd terms. Since the load current has only a fundamental component, only the coefficients $b_1^{(3(2k-1))}$ are present in the differential current. On the other hand, when the load current presents different harmonics, the differential current will result from the contribution of more coefficients. Figure 5.11 shows the current through C_1 and its spectrum for the second example, when the CAMC feeds a rectifier. This current also presents harmonics of the third and ninth orders. There are no common mode components in this current.

We now consider other examples in which there is no need of a DC voltage source. This is the case of a shunt active filter compensating reactive power. Figure 5.12a shows the phase voltage generated by the CAMC together with the corresponding current, which is 90° displaced from the voltage. Figure 5.12b shows the spectrum of i_a, which has a dominant fundamental component.

Figure 5.13a shows the current through C_2 (i_{C2}) together with its spectrum. The spectrum has odd triple components (3rd to 15th), but there are

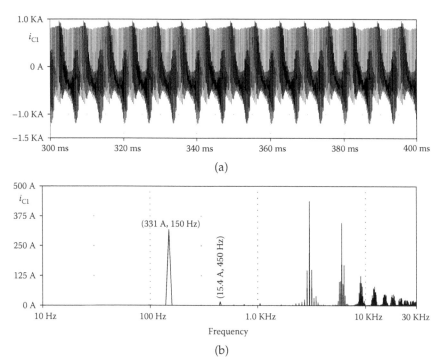

Figure 5.10 CAMC as inverter: (a) current through C_1 and (b) its spectrum.

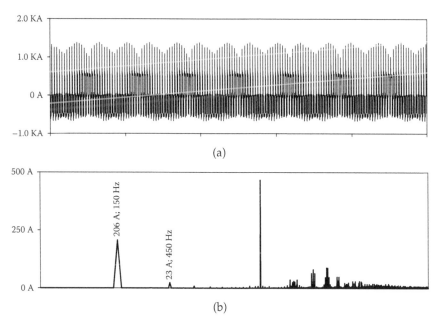

Figure 5.11 (a) Current through C_1 with nonlinear load. (b) Its spectrum.

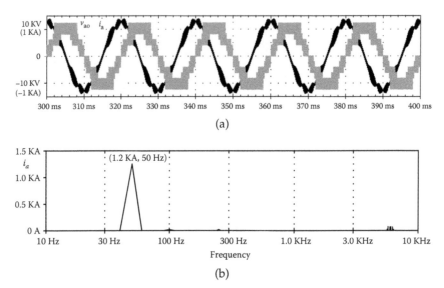

Figure 5.12 CAMC compensating reactive power: (a) phase voltage (2 KV/div) and current (0.2 KA/div) and (b) output current spectrum.

no even components, due to the absence of a common mode current. The amplitude of the third harmonics is 15 times higher than the ninth harmonics, as was shown in the graph of Figure 5.9a. Figure 5.13b shows the current in node M (i_M) together with its spectrum. Comparing the spectrum of both currents, it is observed that i_M has the same harmonics of i_{C2}, but doubles their amplitude.

A last example in this chapter considers the same active filter as before, but now compensating reactive power and harmonics. Figure 5.14 shows the different currents and their spectra. The AC current of the CAMC is shown in Figure 5.14a, where the most significant harmonics are 5th, 7th, 11th, and 13th. Figure 5.14b shows the current in the DC bus, which flows through C_1. Its spectrum presents the common mode harmonics (6th, 12th, and 18th), and the differential harmonics (3rd, 9th, 15th, and 21st) as well. The differential current may be separated looking to the current i_M, which equals the double of the differential current, as shown in Figure 5.14c. The common mode components (6th, 12th, and 18th) are not present in this spectrum.

5.4   Comparison of the Five-Level Topologies

After analyzing several multilevel topologies, a comparison between them can be carried out. In this section, different five-level topologies are compared. These are the most popular: diode-clamped multilevel converter (DCMC), flying capicator multilevel converter (FCMC), cascaded cell multilevel converter (CCMC), and also the hybrid topology presented in this chapter, the CAMC.

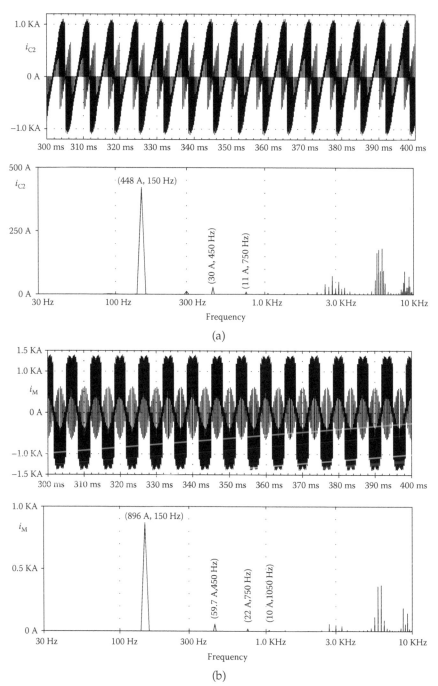

Figure 5.13 Reactive power compensation: (a) current on the DC bus and its spectrum and (b) current in node *M* and its spectrum.

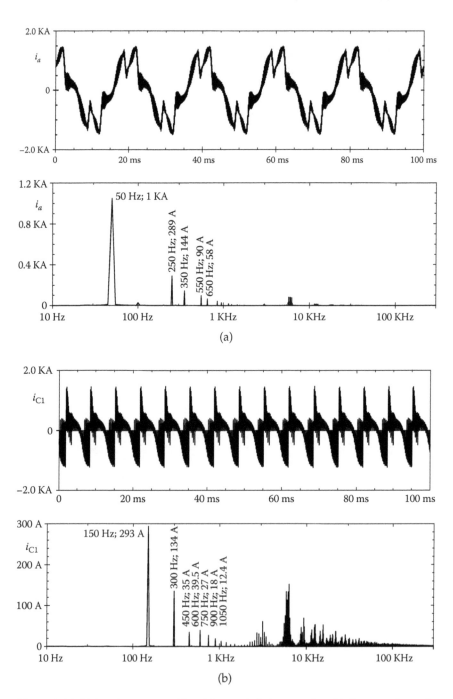

Figure 5.14 (a) Load current with odd nontriple harmonics. (b) Current through capacitor C_1. (c) Differential mode current.

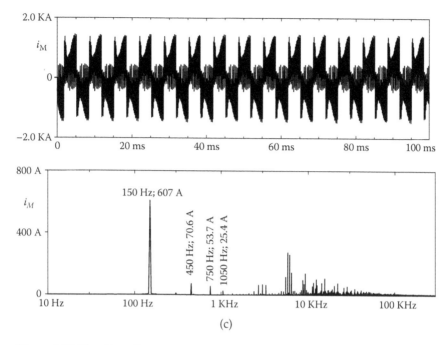

Figure 5.14 (Continued)

5.4.1   DCMC

The five-level DCMC requires 18 clamping diodes. However, some of these diodes should block up to three times the voltage blocked by the power switches. Therefore, if we consider equal blocking voltages for all clamping diodes, this number rises to 36. The conduction interval of the different power switches is not equal. Those near the center of the leg, thus near the load, have longer conduction intervals than those near the positive or negative rails of the DC bus [10]. Another problem to be solved in the DCMC is the voltage balance of the capacitors in the DC bus [4,11].

5.4.2   FCMC

The FCMC requires many capacitors that increase the weight of the converter. A five-level FCMC requires 12 flying capacitors, 3 in each leg of the converter, with different rated voltages. For this, it is difficult to encounter implementations with more than four levels [6,12].

5.4.3   CCMC

Even though we did not analyze the cascaded cell multilevel converter, we are considering it for the comparison since this is an interesting alternative when a common DC bus is not required. The main drawback of this topology is the need for isolated voltage sources for each stage of the converter. Then, it is not possible to use it in back-to-back connection. On the other hand, its modularity provides an easy way to increment the number of voltage levels. It has been applied to reactive compensation where no actual DC sources are needed [13].

5.4.4   CAMC

The CAMC is a hybrid topology that combines the benefits of the DCMC (low number of capacitors) with those of the FCMC (easiness to keep voltage balance). At the same time, it avoids their main drawbacks, like clamping diodes, balancing mechanism, and high number of capacitors.

Table 5.1 summarizes the main characteristics of the four five-level converters. Row 5 shows an energy index that is very important to compare the different topologies regarding the energy storage requirements [9]. It provides a measure of the number of capacitors and their capacitance values, the volume and the cost of the whole converter for a given rated power. The DCMC and the CAMC present the lowest energy index; they are followed by the FCMC, and the CCMC presents a much higher index.

Table 5.1 shows that the CAMC requires fewer components than other topologies. Its energy storage requirement lies between that of the DCMC and the FCMC. It is easier to preserve the voltage balance on the capacitors than in a DCMC. The CAMC is an attractive converter to compete with the DCMC in several medium-voltage applications, such as power conditioning devices like a flexible AC transmission system (FACTS) or custom power devices (CPDs) and electric drives.

Table 5.1 Main Characteristics of the Five-Level Converters

	DCMC	FCMC	CCMC	CAMC
Controlled power switches	24	24	24	24
Clamping diodes	18 or 36	0	0	0
Capacitors in the DC bus	4	1	6	2
Flying capacitors	0	9	0	3
Energy index[a]	2.2	2.9[b]	10	2.5[b]
Voltage balance mechanism	Assisted balance	Natural balance	Natural balance	Natural balance
Back-to-back connection	Yes	Yes	No	Yes

[a] Ripple factor 10%, positive sequence currents.
[b] At f_s = 1 KHz.

5.5 Summary

A new multilevel converter, the cascade asymmetric multilevel converter (CAMC), was introduced in this chapter. It is a five-level converter controlled with a hybrid modulation. It has characteristics similar to those of the HMC, but it uses a single-voltage source for the whole converter. A high-voltage stage, switching at line frequency, is followed by a flying capacitor low-voltage stage, switching at high frequency controlled by a PSPWM. This combination allows obtaining a five-level converter with a simple control. The combination of different topologies leads to a five-level converter with less capacitors than the FCMC counterpart and less diodes than the DCMC.

A detailed study of the low-order harmonics in the DC bus currents has been carried out using a steady-state averaged model. The analysis showed that there exist two different currents in the DC bus. A common mode component whose DC value fixes the active power transfer and its harmonics determine the DC bus ripple. Moreover, there exists a differential component with no DC value, but whose harmonics establish a differential ripple on the DC bus capacitors C_1 and C_2. These harmonics are odd triple harmonics of the modulating frequency. These harmonics and the desired ripple determine the size of the capacitances and the level required for the stored energy in the converter.

References

1. J. Rodríguez, S. Bernet, B. Wu, J.O. Pontt, and S. Kouro. Multilevel Voltage-Source-Converter Topologies for Industrial Medium-Voltage Drives. *IEEE Transactions on Industrial Electronics*, 54(6), 2930–2945, 2007.
2. S.A. González, M.I. Valla, C.F. Christiansen. Analysis of a Cascade Asymmetric Topology for Multilevel Converters. In *IEEE International Symposium on Industrial Electronics (ISIE 2007)*, Vigo, Spain, June 4–7, 2007, pp. 1027–1032.
3. S. Kouro, M. Malinowski, K. Gopakumar, J. Pou, L.G. Franquelo, B. Wu, J. Rodriguez, M.A. Pérez, J.I. Leon. Recent Advances and Industrial Applications of Multilevel Converters. *IEEE Transactions on Industrial Electronics*, 57(8), 2553–2580, 2010.
4. H. Abu-Rub, J. Holtz, J. Rodriguez, G. Baoming. Medium-Voltage Multilevel Converters—State of the Art, Challenges, and Requirements in Industrial Applications. *IEEE Transactions on Industrial Electronics*, 57(8), 2581–2596, 2010.
5. ABB Switzerland Ltd. Medium Voltage Drives. www.abb.com/drives.
6. Alstom Power Conversion. Drive Range Drive Solutions for All Applications. www.alstom.com.
7. ABB, Inc. Power Products and Power Systems, Battery Energy Storage Optimizes Integration of Renewable Energy to the Grid. www.abb.us/smartgrids.
8. Why Is Battery Energy Storage Such a Hot Topic? A North American Perspective on Smart Grid Trends and Technologies. *ABB Smart Grids*, January 6, 2012.

9. S.A. González. Multilevel Converters: Their Application in Medium Voltage Power Systems (in Spanish). PhD thesis, School of Engineering, National University of La Plata, Argentina, December 2010.

10. T. Brückner, S. Bernet, H. Güldner. The Active NPC Converter and Its Loss-Balancing Control. *IEEE Transactions on Industrial Electronics*, 52(3), 855–868, 2005.

11. M. Saeedifard, R. Iravani, J. Pou. Analysis and Control of DC-Capacitor-Voltage-Drift Phenomenon of a Passive Front-End Five-Level Converter. *IEEE Transactions on Industrial Electronics*, 54(6), 3255–3266, 2007.

12. D. Krug, S. Bernet, S.S. Fazel, K. Jalili, M. Malinowski. Comparison of 2.3-kV Medium-Voltage Multilevel Converters for Industrial Medium-Voltage Drives. *IEEE Transactions on Industrial Electronics*, 54(6), 2979–2992, 2007.

13. F. Peng, J. Wang. A Universal STATCOM with Delta-Connected Cascade Multilevel Inverter. In *IEEE Power Electronics Specialists Conference (PESC'04)*, Aachen, Germany, June 20–25, 2004, pp. 3529–3533.

chapter 6

Case Study 1: DSTATCOM Built with a Cascade Asymmetric Multilevel Converter

6.1 Introduction

The power quality problems found in the distribution network have different causes [1]. Most of them are originated by the consumers, such as electric arc furnaces that cause flicker [2], nonlinear loads that introduce harmonic currents [3], unbalanced loads, and connection and disconnection of electric motors in high-power industries that cause voltage sags [4.]

Low power factor and harmonic currents are the most common power quality problems in medium-voltage (MV) distribution systems. One way to compensate these disturbances is to employ custom power devices (CPDs), such as distribution static compensator (DSTATCOM), which is mainly used for reactive power compensation [5,6] and active and hybrid filters for the compensation of harmonic currents [7,8]. The voltage source multilevel converters are a good choice to implement CPDs in MV without using coupling transformers [9]. In particular, the cascade asymmetric multilevel converter (CAMC) presented in Chapter 5 is very attractive to implement a CPD. A DSTATCOM implemented with a CAMC is presented in this chapter. It is used to compensate reactive power and harmonics at the point of common coupling (PCC).

In the next section a dynamic model of the CAMC is obtained, in order to be used in the control of the active and reactive currents of the converter. The model is developed in a synchronous reference frame *q-d-0*, and it is based on the instantaneous reactive power theory [10]. Then it is possible to design two decoupled loops that control the active and reactive power consumed or generated by the CAMC. Afterwards, the CAMC is connected to the PCC through a coupling inductor to compensate linear and nonlinear loads.

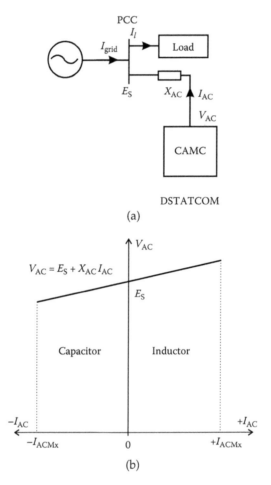

Figure 6.1 (a) Connection scheme of the DSTATCOM and the CAMC. (b) *V-I* characteristics of the DSTATCOM.

6.2 Compensation Principles

Figure 6.1a presents a simple connection scheme of a DSTATCOM at the PCC. The medium-voltage system is represented by an infinite power source E_S, a passive load (Z_L), and the CAMC, which is coupled to the PCC through an inductor (X_{AC}). The combination of the CAMC and the X_{AC} constitutes a DSTATCOM. The amount of current delivered by the DSTATCOM (I_{AC}) is determined by the voltage at the PCC (E_S), the voltage generated by the converter (V_{AC}), and the coupling impedance:

$$I_{AC} = \frac{V_{AC} - E_S}{X_{AC}} \tag{6.1}$$

The voltage difference is applied on the inductor, fixing the current I_{AC}. Assuming that the voltage is imposed by the infinite power source, the reactive power delivered by the DSTATCOM is given by

$$Q_{AC} = E_S \cdot I_{AC} \qquad (6.2)$$

Then, the reactive power control is carried out regulating the amplitude of V_{AC}. On the other hand, its relative phase shift with respect to the system voltage is controlled to compensate the power losses of the converter.

Figure 6.1b shows the *V-I* output characteristic of the DSTATCOM. It presents a regulation slope that is fixed by the coupling reactance. The voltage for zero current corresponds to the rated voltage at the PCC. The characteristics are limited to the maximum current that can be delivered by the converter ($\pm I_{ACMx}$). When the load consumes lagging current, the compensator should behave as a capacitor, and it should look like an inductor when the load consumes leading current.

The phasor diagrams of Figure 6.2 show the steady-state compensation characteristics of the CAMC for both situations, inductive or capacitive

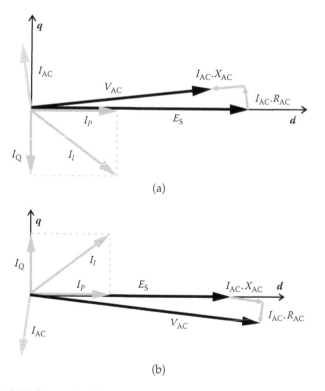

(a)

(b)

Figure 6.2 (a) Inductive load compensation. (b) Capacitive load compensation.

load. In these diagrams the losses in the converter are represented with a resistance (R_{AC}) connected in series with the coupling inductor. Figure 6.2a corresponds to an inductive load, in which the load current (I_l) is delayed with respect to the system voltage E_S. The reactive power demanded by the load is determined by the current I_Q, which is 90° delayed with respect to E_S. A total compensation is obtained when the CAMC injects a current with the same amplitude and the opposite phase of I_Q. The actual current I_{AC} has a small component in phase with E_S in order to compensate the losses in R_{AC}. This is obtained controlling the output voltage of the CAMC, which should be in advance to the system voltage and with amplitude slightly smaller than E_S. Figure 6.2b corresponds to the case of a capacitive load. The load current I_l now is in advance to E_S, and the CAMC should look like an inductor. Then, V_{AC} is delayed and slightly higher than the PCC voltage. Again, the current I_{AC} has a small component in phase with E_S to compensate the converter losses.

6.3   CAMC Model

The dynamic behavior of a DSTATCOM implemented with a CAMC may be obtained analyzing the circuit presented in Figure 6.3. Here the CAMC is connected to the PCC through L_{AC} and R_{AC}. The coupling inductor has a double function: On one side it separates the instantaneous values of the system voltage and that generated by the CAMC. On the other side it determines the steady-state current delivered by the DSTATCOM, as was shown in the previous section. It was found in Chapter 5 that the main component of the output voltage of the CAMC corresponds to its average value over one switching period. Thus, for a balanced load it results in

$$\bar{v}_{ao} = m\frac{V_{DC}}{2}\sin\left(\omega_m t\right)$$

$$\bar{v}_{bo} = m\frac{V_{DC}}{2}\sin\left(\omega_m t - \frac{2}{3}\pi\right) \tag{6.3}$$

$$\bar{v}_{co} = m\frac{V_{DC}}{2}\sin\left(\omega_m t + \frac{2}{3}\pi\right)$$

Considering slow time variations, when compared with the switching period on the DC voltage and on the modulation index, these variations produce slow changes on the phase voltage amplitude, so they are indicated with the symbol ~. Then they may be represented in matrix form:

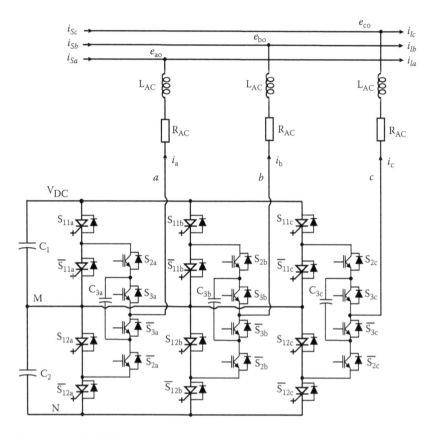

Figure 6.3 DSTATCOM built with a CAMC.

$$\begin{bmatrix} \tilde{v}_{ao} \\ \tilde{v}_{bo} \\ \tilde{v}_{co} \end{bmatrix} = \tilde{m}\,\frac{\tilde{v}_{DC}}{2}\,[M]^{T} \tag{6.4}$$

where

$$[M]^{T} = \begin{bmatrix} \sin(\omega{\cdot}t) \\ \sin(\omega{\cdot}t - \tfrac{2}{3}\pi) \\ \sin(\omega{\cdot}t + \tfrac{2}{3}\pi) \end{bmatrix}$$

The control signals for each phase are defined as

$$
\begin{bmatrix} \tilde{m}_a \\ \tilde{m}_b \\ \tilde{m}_c \end{bmatrix} = \tilde{m} \cdot \begin{bmatrix} \sin(\omega \cdot t) \\ \sin(\omega \cdot t - \tfrac{2}{3}\pi) \\ \sin(\omega \cdot t + \tfrac{2}{3}\pi) \end{bmatrix} = \tilde{m} \cdot \begin{bmatrix} M \end{bmatrix}^T
\tag{6.5}
$$

Finally, (6.4) may be written as

$$
\begin{bmatrix} \tilde{v}_{ao} \\ \tilde{v}_{bo} \\ \tilde{v}_{co} \end{bmatrix} = \frac{\tilde{v}_{DC}}{2} \begin{bmatrix} \tilde{m}_a \\ \tilde{m}_b \\ \tilde{m}_c \end{bmatrix}
\tag{6.6}
$$

The dynamic model of the CAMC connected to an infinite power source, in phase variables, is represented by the following matrix equation:

$$
\begin{bmatrix} \dfrac{d\tilde{i}_a}{dt} \\[2mm] \dfrac{d\tilde{i}_b}{dt} \\[2mm] \dfrac{d\tilde{i}_c}{dt} \end{bmatrix} = -\frac{R_{AC}}{L_{AC}} \begin{bmatrix} \tilde{i}_a \\ \tilde{i}_b \\ \tilde{i}_c \end{bmatrix} - \frac{1}{L_{AC}} \begin{bmatrix} e_{ao} \\ e_{bo} \\ e_{co} \end{bmatrix} + \frac{1}{L_{AC}} \frac{\tilde{v}_{DC}}{2} \begin{bmatrix} \tilde{m}_a \\ \tilde{m}_b \\ \tilde{m}_c \end{bmatrix}
\tag{6.7}
$$

The vector e_{io} corresponds to the system voltage at the PCC.

The dynamic behavior of the DC bus was analyzed in Chapter 5. Figure 6.4 reproduces the circuit representation of the DC bus. It has three current generators representing the differential mode currents $\tilde{i}_{M/2}$, which are applied to each capacitor in the bus, and the common mode current $\tilde{i}_C$, which is applied to both capacitors. Different from Chapter 5, here a resistor (R_C) is included to represent the losses on the DC side of the converter. Considering that both capacitors have the same capacitance C, the DC voltage variation is proportional to the sum of the different currents through each capacitor:

$$
\frac{d\tilde{v}_{DC}}{dt} = \frac{1}{C} \left(\tilde{i}_{C1} + \tilde{i}_{C2} \right)
\tag{6.8}
$$

These currents depend on the common and differential mode generators:

$$
\tilde{i}_{C1} = \frac{\tilde{i}_M}{2} + \tilde{i}_C - \frac{\tilde{v}_{DC}}{R_C} \quad \text{and} \quad \tilde{i}_{C2} = -\frac{\tilde{i}_M}{2} + \tilde{i}_C - \frac{\tilde{v}_{DC}}{R_C}
\tag{6.9}
$$

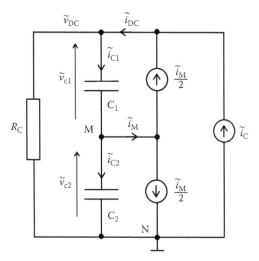

Figure 6.4 Dynamic model of the DC bus.

Then, (6.8) results in

$$\frac{d\tilde{v}_{DC}}{dt} = \frac{2}{C}\left(\tilde{i}_C - \frac{\tilde{v}_{DC}}{R_C}\right) \tag{6.10}$$

The common mode current was obtained in Chapter 5. Its averaged value can be expressed as

$$\tilde{i}_{DC} = -\left(s_{1a}\tilde{i}_a\tilde{m}\sin\left(\omega_m t\right) + s_{1b}\tilde{i}_b\tilde{m}\sin\left(\omega_m t - \tfrac{2}{3}\pi\right) + s_{1c}\tilde{i}_c\tilde{m}\sin\left(\omega_m t + \tfrac{2}{3}\pi\right)\right) \tag{6.11}$$

Replacing the switching functions by their Fourier series as described in (5.15), and considering the averaged value of this series, the averaged value of the DC bus current is obtained taking into account (6.5):

$$\tilde{i}_C = \frac{1}{2}\begin{bmatrix} \tilde{m}_a & \tilde{m}_b & \tilde{m}_c \end{bmatrix}\begin{bmatrix} \tilde{i}_a \\ \tilde{i}_b \\ \tilde{i}_c \end{bmatrix} \tag{6.12}$$

Replacing (6.12) in (6.10), the dynamic model of the DC bus is obtained:

$$\frac{d\tilde{v}_{DC}}{dt} = -\frac{1}{C}\begin{bmatrix} \tilde{m}_a & \tilde{m}_b & \tilde{m}_c \end{bmatrix}\begin{bmatrix} \tilde{i}_a \\ \tilde{i}_b \\ \tilde{i}_c \end{bmatrix} - \frac{2\tilde{v}_{DC}}{R_C C} \tag{6.13}$$

The complete dynamic model of the DSTATCOM, with $\tilde{m}_i$ as control input, is represented by (6.7) and (6.13).

It is well known that a three-phase system in phase coordinates (*abc*) is strongly coupled, so it is more convenient to transform the representation into a synchronous reference frame (*qd0*) in order to properly design the control of the DSTATCOM. The coordinate transformation is done with the matrix **K**:

$$K = \frac{2}{3}\begin{bmatrix} \cos\theta & \cos(\theta - \frac{2}{3}\pi) & \cos(\theta + \frac{2}{3}\pi) \\ \sin\theta & \sin(\theta - \frac{2}{3}\pi) & \sin(\theta + \frac{2}{3}\pi) \\ \frac{1}{2} & \frac{1}{2} & \frac{1}{2} \end{bmatrix} \tag{6.14}$$

where $\theta = \int \omega(\tau)\cdot d\tau + \theta_0 = \omega_0 t + \theta_0$, and θ_0 is the initial phase displacement of the phase voltage e_{ao}, with respect to the reference frame. In this way the **d-axis** in the synchronous reference frame is aligned with phase **a** of the three-phase system:

$$e_{ao} = \sqrt{\frac{2}{3}} E_{ll(rms)} \sin\omega_0 t \tag{6.15}$$

The dynamic model represented in the synchronous reference frame for a three-wire system results in

$$\frac{d}{dt}\begin{bmatrix} \tilde{i}_q \\ \tilde{i}_d \end{bmatrix} = \begin{bmatrix} -\dfrac{R_{AC}}{L_{AC}} & -\omega \\ \omega & -\dfrac{R_{AC}}{L_{AC}} \end{bmatrix}\cdot\begin{bmatrix} \tilde{i}_q \\ \tilde{i}_d \end{bmatrix} - \frac{1}{L_{AC}}\begin{bmatrix} 0 \\ e_d \end{bmatrix} + \frac{\tilde{v}_{DC}}{2L_{AC}}\begin{bmatrix} \tilde{m}_q \\ \tilde{m}_d \end{bmatrix} \tag{6.16a}$$

$$\frac{d\tilde{v}_{DC}}{dt} = -\frac{3}{C}\begin{bmatrix} \tilde{m}_q & \tilde{m}_d \end{bmatrix}\begin{bmatrix} \tilde{i}_q \\ \tilde{i}_d \end{bmatrix} - \frac{2\tilde{v}_{DC}}{R_C C} \tag{6.16b}$$

Then, the whole converter model is represented by (6.16a) and (6.16b).

6.3.1 Current Control

The active and reactive power managed by the CAMC is calculated using the instantaneous reactive power theory [10]. Then the instantaneous power of the CAMC is defined as

$$p_e = e_{do}\tilde{i}_d + e_{qo}\tilde{i}_q$$
$$q_e = -e_{do}\tilde{i}_q + e_{qo}\tilde{i}_d \tag{6.17}$$

Remembering that the d-axis is aligned with the system phase a, the voltage component along the q-axis is null. So the power expression is simplified to

$$p_e = e_{do}\tilde{i}_d$$
$$q_e = -e_{do}\tilde{i}_q \tag{6.18}$$

In this way, the active power depends only on the current along the d-axis ($\tilde{i}_d$), while the reactive power depends on the current along the q-axis ($\tilde{i}_q$).

Moreover, assuming that the DC bus voltage has a negligible ripple and that its average value is kept constant with a control loop, $\tilde{v}_{DC}$ is constant and equal to its reference value. Under these conditions the AC side model of the DSTATCOM, given in (6.16a), can be expressed as

$$L_{AC}\frac{d}{dt}\begin{bmatrix}\tilde{i}_q\\\tilde{i}_d\end{bmatrix}+R_{AC}\begin{bmatrix}\tilde{i}_q\\\tilde{i}_d\end{bmatrix}=\begin{bmatrix}0 & -\omega_m L_{AC}\\\omega_m L_{AC} & 0\end{bmatrix}\begin{bmatrix}\tilde{i}_q\\\tilde{i}_d\end{bmatrix}-\begin{bmatrix}0\\e_{do}\end{bmatrix}+\frac{V_{DC}}{2}\begin{bmatrix}\tilde{m}_q\\\tilde{m}_d\end{bmatrix} \tag{6.19}$$

The first term of the right side of the equation shows that there is some coupling between both axes when the voltages are the control inputs. It is desirable to have two independent controls for the active and the reactive power, so this coupling should be compensated. This is possible with a feed-forward compensation that introduces these terms with the same value and opposite sign in each control loop, based on the phase current measurements [9,11].

Figure 6.5 shows the control scheme for both loops; they are based on simple proportional plus integral (PI) controllers. Here it is possible to see the feed-forward terms $\omega_m L_{AC}\cdot i_{dm}$ in the **q-axis** control loop and $-\omega_m L_{AC}\cdot i_{qm}$ in the **d-axis** control loop. Moreover, the voltage at the PCC (E_S) is also added up in the **d-axis** control loop to diminish the control efforts of the actuator.

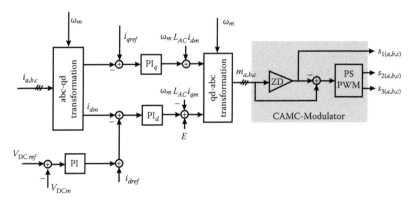

Figure 6.5 Block diagram for the current control and the modulator of the CAMC.

The outputs of these controllers ($\tilde{m}_{d,q}$) are the modulation references for the voltages to be generated by the CAMC. These signals are transformed to phase variables ($\tilde{m}_{a,b,c}$) that are the inputs to three zero detectors in Figure 6.5. The output of these detectors are square waveforms at the modulation frequency and constitute the switching signals $s_{1(a,b,c)}$, which control the high-voltage (HV) stages of each leg of the CAMC. At the same time, the square signals $s_{1(a,b,c)}$ are subtracted from the reference values $\tilde{m}_{a,b,c}$ to obtain the modulation signals for the low-voltage (LV) stages of the three legs of the CAMC. The modulation of the LV stages uses phase-shifted carrier PWM for each leg of the converter, generating the switching signals $s_{2(a,b,c)}$ and $s_{3(a,b,c)}$.

Figure 6.5 also shows the DC bus voltage control loop. This control defines the active power requirements (mainly due to R_{AC} and R_C) and commands the current along the *d-axis*.

Figure 6.6 shows the different waveforms generated by the CAMC when it is used for reactive compensation of an inductive load. Figure 6.6a shows the reactive power injected by the CAMC to the load when the load demand increases from 10 MVA to 20 MVA. This change produces an increment of the converter current, as predicted by (6.1) and (6.2), and it is shown in Figure 6.6b for phase *a* current (i_a). The phase voltage is also shown here to see that the phase displacement of both signals remains constant and equal to 90°. The current has a leading phase angle with respect to the phase voltage v_{ao}. Figure 6.6c shows the DC bus voltage, which has a transient drop due to the sudden increment of the delivered current. It is clear that the DC bus supplies the energy required for the change in current amplitude, but the control loops restore it to the reference value after 40 ms. The increment of the converter currents also increases the power losses, which increase the active power drawn from the PCC as shown in Figure 6.6d.

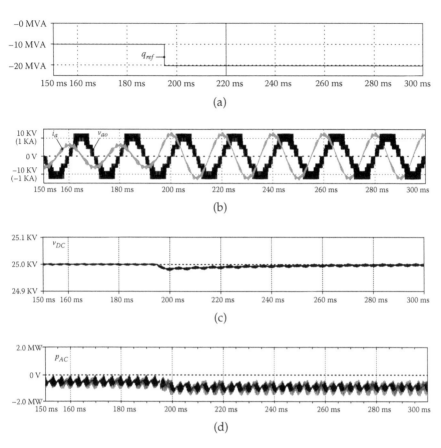

Figure 6.6 Reactive compensation: Q from −10 MVA to −20 MVA: (a) reactive power reference, (b) output current and phase voltage of the CAMC, (c) DC bus voltage, and (d) active power.

Figure 6.7 shows the results of a load change from inductive to capacitive, with the same power level. So the CAMC should change the kind of compensation from a leading current to lagging current. Figure 6.7a shows the reactive power injected by the CAMC to the load, when the load demand varies from −20 MVA to 20 MVA. This change produces a 180° phase shift in the converter currents, preserving their amplitude, as shown in Figure 6.7b, for phase *a* current (i_a). Figure 6.7c shows the phase voltage v_{ao} generated by the CAMC. It is clear that the phase displacement between phase voltage and current changes from +π/2 to −π/2. Figure 6.7d shows the disturbances in the DC bus voltage, generated by the load demand. The phase change of the generated current provokes a current spike that increments the DC bus voltage, but the control loop restores the reference value after 40 ms.

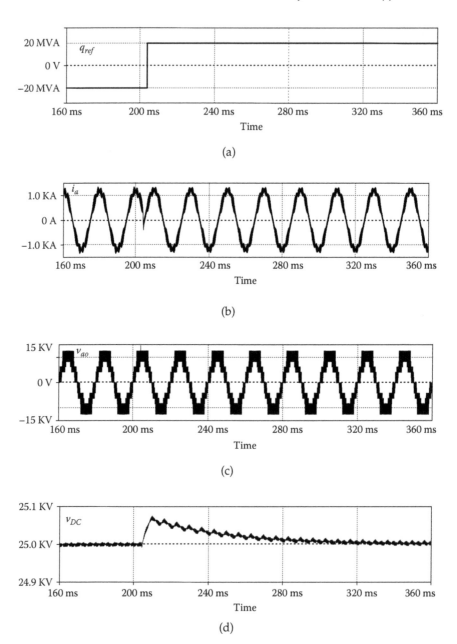

Figure 6.7 Reactive compensation: Q from −20 MVA to 20 MVA: (a) reactive power reference, (b) phase current of the CAMC, (c) phase voltage of the CAMC, and (d) DC bus voltage.

6.4 Reactive Power and Harmonics Compensation

Another example of the CAMC used as a custom power device is illustrated in Figure 6.8. A medium-voltage power system of 13.8 kV and a short-circuit power of 3 GVA feeds a variable linear load and a nonlinear load. The CAMC is connected to the PCC through an inductor. The nonlinear load is a diode three-phase bridge rectifier that consumes 18 MW and is connected through an inductance L_r (0.7 mH). The linear load is inductive and varies between 20 and 43 MVA with a power factor equal to 0.86 and 0.7, respectively. The rated power for the CAMC is fixed to 20 MVA to guarantee that the power factor seen by the system at the PCC is always higher than 0.9, and that the system current has a low THD.

The reactive power and harmonics compensator provides the reactive current and harmonics demanded by the load so that the distribution network does not see these disturbances. The compensation is performed through two cascaded control loops. The outer loops calculate the references to the inner current control loops from the measurements at the PCC. The inner control loops, shown in gray in Figure 6.8, correspond to the current control of the CAMC, as discussed in the previous section. Each loop corresponds to the d-axis or q-axis current. The DC bus voltage control determines the current in d-axis i_{dref}. The current along the q-axis, i_{qref} is mainly determined by the reactive current required by the load. The measurement of the load current is of outmost importance to prevent the circulation of reactive current and harmonics through the network.

6.4.1 System Model

It is important to know the nature of the load in order to compensate the disturbances introduced by the load through a shunt compensator. The load current (i_l) may be expressed as the sum of the current provided by the network (i_S), plus the compensation current introduced by the CAMC (i_{AC}). Then,

$$i_l = i_S + i_{AC} \tag{6.20}$$

In a synchronous reference frame (q-d) they are

$$\begin{bmatrix} i_{lq} \\ i_{ld} \end{bmatrix} = \begin{bmatrix} i_{Sq} + i_q \\ i_{Sd} + i_d \end{bmatrix} \tag{6.21}$$

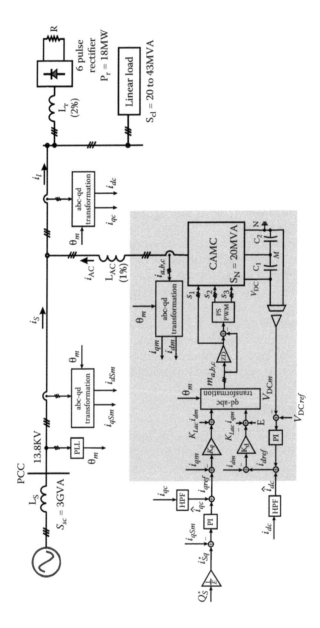

Figure 6.8 Scheme of the compensating system at the PCC.

where

$$i_{AC} = \begin{bmatrix} i_q \\ i_d \end{bmatrix}, \quad i_l = \begin{bmatrix} i_{lq} \\ i_{ld} \end{bmatrix} \quad \text{and} \quad i_S = \begin{bmatrix} i_{Sq} \\ i_{Sd} \end{bmatrix}$$

i_{AC} should be such that i_S has very low harmonics distortion and it is almost in phase with the voltage at the PCC. Taking into account that the voltage at the PCC has a null component along the q-axis, the power supplied by the network can be written as

$$q_S = -e_{do} \cdot i_{Sq} = -e_{do}\left(i_{lq} - i_q\right)$$
$$p_S = e_{do} \cdot i_{Sd} = e_{do}\left(i_{ld} - i_d\right)$$

(6.22)

Then, the CAMC should provide the following currents:

$$i_q = i_{lq} - i_{Sq}^*$$
$$i_d = i_{ld} - i_{Sd}^*$$

(6.23)

where i_{Sq}^* and i_{Sd}^* are the reference values of the network currents. Generally, $i_{Sq}^* = 0$ when the reactive power is fully compensated. Nevertheless, it may take a different value when a power factor lower than 1 is accepted. On the other hand, the active current reference (i_{Sd}^*) should be such that i_d equals zero. But, there is no control of the average value of the current along the d-axis; i_{Sd} will provide the active power to the load plus the losses of the CAMC.

6.4.2 Reactive Power Compensation

First, considering only the linear load of the system shown in Figure 6.8, the reactive power control at the PCC can be reduced to the schemes shown in Figure 6.9, for each one of the axes d and q.

The gray area in Figure 6.8 indicates the model of the CAMC obtained when both currents $\tilde{i}_q$ and $\tilde{i}_d$ are controlled (Figure 6.5). From (6.19) and doing an adequate design of the controller, the CAMC plus current control can be modeled as a first-order system, whose time constants are given by

$$\tau_d = \frac{\tau_{AC}}{\left(1 + {K_d}/{R_{AC}}\right)} \quad \text{and} \quad \tau_q = \frac{\tau_{AC}}{\left(1 + {K_q}/{R_{AC}}\right)}$$

(6.24)

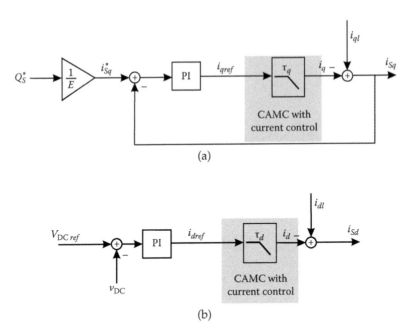

Figure 6.9 Reactive power control loops: (a) *q*-axis control loop and (b) *d*-axis control loop.

where $\tau_{AC} = L_{AC}/R_{AC}$, K_q, and K_d are the gains of current controllers.

Figure 6.9a shows the external control loop, which fixes the reference current along the *q*-axis to control the reactive power provided by the network as indicated in (6.22). Figure 6.9b corresponds to the active power control loop, which is fixed by the DC voltage control.

Figure 6.10 shows the simulation waveforms of phase *a* corresponding to the CAMC compensating the reactive power of a balanced linear load with an apparent power of 32 MVA and an inductive cos φ = 0.86. The CAMC starts the compensation at *t* = 150 ms, providing the reactive power required by the load. Figure 6.10a shows the three currents at the PCC (i_{Sa}, i_{la}, and i_a). Before the CAMC starts the compensation, the network current equals the load current. After that, the system current reduces its amplitude and presents a phase shift toward alignment with the phase voltage at the PCC, as can be seen in Figure 6.10b. The improvement of the power factor seen by the network becomes clear when the CAMC starts the compensation.

Another way to evaluate the improvement in the power factor is defining an index that relates the instantaneous active power with the whole instantaneous power. This is named instantaneous power index (*ipi*) and is defined as

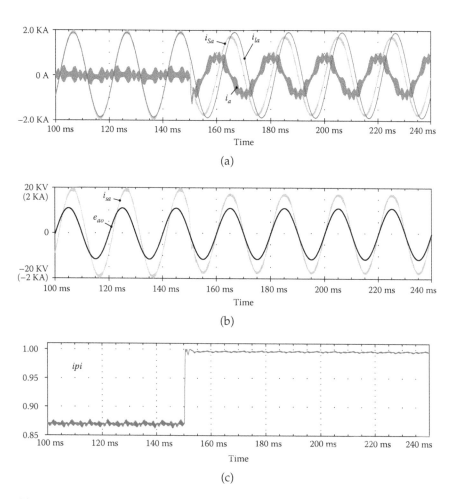

Figure 6.10 Reactive power compensation: (a) compensator, system, and load currents, i_a, i_{Sa}, and i_{la}; (b) phase voltage and current at the PCC; and (c) instantaneous power index at the PCC.

$$ipi = \frac{p}{\sqrt{p^2 + q^2}} \tag{6.25}$$

While working in steady state and sinusoidal waveforms, the *ipi* equals the power factor defined by cos φ. Figure 6.10c shows the *ipi* at the PCC for the presented test. While the CAMC is not working, the *ipi* equals the load power factor, which is inductive and equal to 0.87. Afterwards the *ipi* increases up to a value near unity.

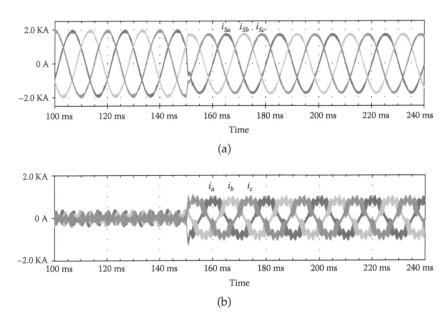

Figure 6.11 (a) Line currents at the PCC. (b) Currents delivered by the CAMC.

Figure 6.11 shows the three-phase currents of the network and the CAMC. It is clear that the behavior described for phase *a* is reproduced in the other two-phase currents.

Another example shown here is the capability of DSTATCOM to compensate the load variations related to the required active and reactive power. DSTATCOM allows maintaining the network power factor near unity regardless of the load. In this test a balanced linear load varies from its initial value of apparent power of 32 MVA with an inductive cos φ = 0.86 to a final value of 42.4 MVA with an inductive cos φ = 0.7. Figure 6.12 shows the simulation waveforms of phase *a* for a sudden change of the load. The different currents flowing to the PCC (i_{Sa}, i_{la}, and i_a) are compared in Figure 6.12a. The CAMC controls the reactive power at the PCC so that the network current (i_{Sa}) is almost in phase with the network voltage (e_{ao}), as can be seen in Figure 6.12b. The only effect observed at the PCC is the increment of the three currents in phase with the PCC voltage. Figure 6.12c shows that the CAMC current is almost 90° in advance of the voltage generated, consuming a negligible active power while providing the reactive power to the load.

6.4.3 Harmonics Current Compensation

The presence of nonlinear loads introduces harmonic currents that disturb the network voltage. One of the most common nonlinear loads

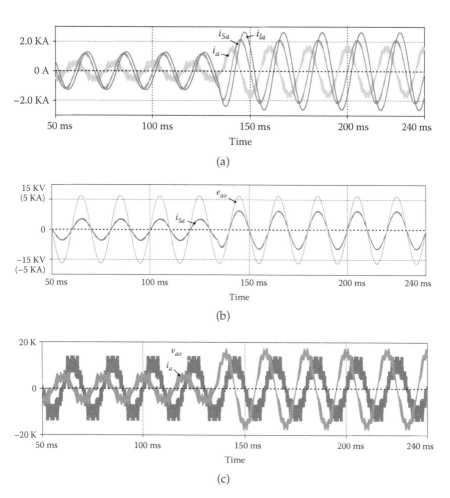

Figure 6.12 Load change from 20 MVA with PF = 0.86 to 43.3 MVA with PF = 0.7: (a) current waveforms i_a, i_{Sa}, and i_{la}; (b) voltage and current at the PCC; and (c) voltage and current generated by the CAMC.

found in industry is the three-phase bridge rectifier, which introduces odd harmonics in the load current. This kind of load does not introduce zero-sequence harmonics. It can be compensated with passive filters, active power filters, or a combination of both as a hybrid power filter [8]. In the system shown in Figure 6.8, the CAMC is used to compensate the harmonics required by the six-pulse rectifier connected to the PCC. The CAMC provides the harmonics that are generated by the load so that the network current remains almost free of harmonic distortion. Coming back to (6.23), the load current, in a synchronous reference frame, may be separated in two terms: a DC value ($\bar{i}_{l(q,d)}$) that corresponds to the

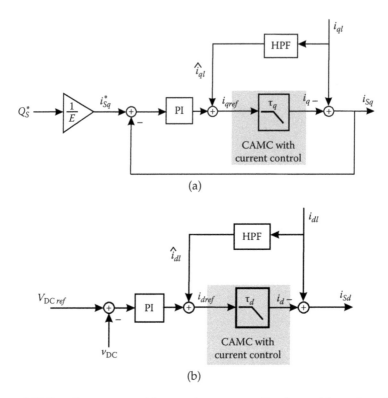

Figure 6.13 Reactive power and harmonics compensation loops: (a) *q*-axis control loop and (b) *d*-axis control loop.

fundamental component and an AC value $(\hat{i}_{l(q,d)})$ that corresponds to the harmonics. Then the current that should be provided by the CAMC is

$$i_q = \left(\tilde{i}_{lq} + \hat{i}_{lq}\right) - i^*_{Sq}$$

$$i_d = \hat{i}_{lq}$$

(6.26)

Figure 6.13 shows the control loops for the two-axis *q* and *d*. Here, two high-pass filters (HPFs) are introduced in order to extract the harmonics of the load currents $\hat{i}_{lq}$ and $\hat{i}_{lq'}$ which should be provided by the CAMC.

A new example considers a balanced linear load with an apparent power of 32 MVA and an inductive cos φ = 0.86, together with a six-pulse rectifier that feeds an 18 MW DC load. Figures 6.14 and 6.15 show the simulation results of this example. The different waveforms for phase *a* are presented in Figure 6.14, while the harmonic spectra of the different currents flowing toward the PCC are shown in Figure 6.15.

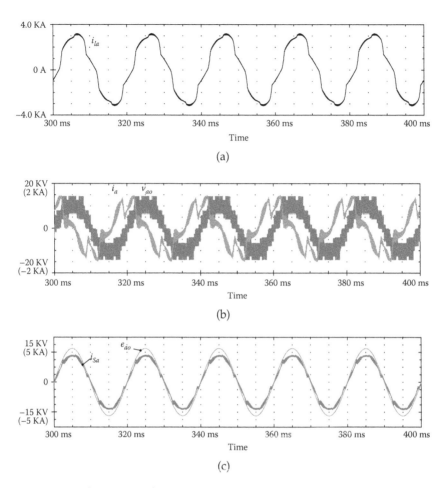

Figure 6.14 Phase *a* waveforms: (a) load current, (b) CAMC current and voltage, and (c) system current and voltage at the PCC.

The load current is shown in Figure 6.14a, while its spectrum is presented in Figure 6.15a. The graphic shows the whole spectrum where the fundamental component prevails over the harmonics introduced by the nonlinear load. A zoom of the graphic is shown to see the details of the harmonic contents. The total harmonic distortion of the load current (THD$_i$) equals 11.12% when the first 11 harmonics are taken into account. The voltage generated by the CAMC, together with the current fed by the active filter, is shown in Figure 6.14b. The current spectrum is presented in Figure 6.15b. The graphic illustrates the fundamental component that corresponds to the reactive current required by the load and which is smaller than the fundamental component of the load and the network current. A zoom of the graphic shows the details of the harmonics

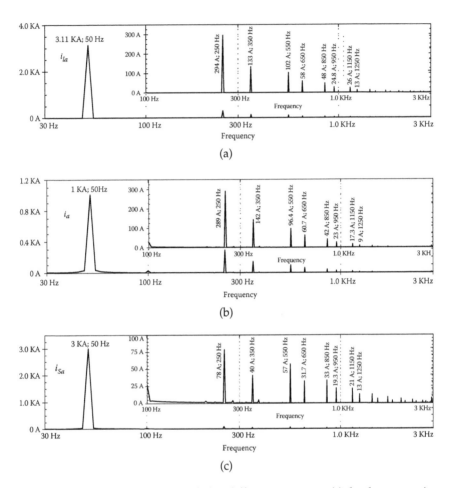

Figure 6.15 Harmonic spectra of the different currents: (a) load current i_{la}, (b) CAMC current i_a, and (c) system current i_{Sa}.

spectrum. It is easy to see that the harmonics are almost equal to those of the load. The network current is shown in Figure 6.14c, together with the phase voltage at the PCC. Figure 6.15c shows the spectrum of the network current. The graphic illustrates the fundamental component that corresponds to the active current required by the load. A zoom of the graphic shows a detail of the harmonics spectrum. It is easy to see that the system harmonics are smaller than those of the load, but they are not completely eliminated. The CAMC is not capable of fully compensating the harmonics, mainly due to the bandwidth, which is limited by the switching frequency. In this case the closed loops shown in Figure 6.13 have a cutoff frequency of 1.33 kHz. The THD_i of the network current (i_S) is 3.98% when the first 11 harmonics are taken into account.

6.5   Summary

A case study of a cascaded asymmetric multilevel converter compensating reactive power and harmonics was presented in this chapter. The full compensation was performed controlling the active and reactive currents in a synchronous reference frame. First, an averaged model of the system in q-d frame is obtained. Then two decoupled current loops are developed so that the reactive power is controlled with the current along the q-axis, while the active power is controlled with the current along the d-axis. The active current only provides the losses of the CAMC. Finally, a nonlinear load was added and the CAMC provided the reactive power plus the harmonics currents. Even when the CAMC has a small bandwidth it is capable of reducing the THD from an original 11.12% to less than 4% in the network current at the PCC.

References

1. M.M. Morcos, J.C. Gomez. Electric Power Quality—The Strong Connection with Power Electronics. *IEEE Power and Energy Magazine*, 1(5), 18–25, 2003.
2. P. Lauttamus, H. Tuusa. Simulated Electric Arc Furnace Voltage Flicker Mitigation with 3-Level Current-Controlled STATCOM. In *IEEE Applied Power Electronics Conference and Exposition (APEC'08)*, February 24–28, 2008, pp. 1697–1703.
3. A. Rahmati, A. Abrishamifar, E. Abiri. An DSTATCOM for Compensating Different Abnormal Line Voltage and Nonlinear Load. In *IEEE International Conference on Industrial Technology (ICIT'06)*, Mumbai, India, December 15–17, 2006, pp. 756–761.
4. B.N. Singh, A. Adya, A.P. Mittal, J.R.P. Gupta, B. Singh, Application of DSTATCOM for Mitigation of Voltage Sag for Motor Loads in Isolated Distribution. In *IEEE International Symposium on Industrial Electronics (ISIE'06)*, Montreal, Canada, July 9–13, 2006, pp. 1806–1811.
5. A. Ghosh, G. Ledwich. *Power Quality Enhancement Using Custom Power Devices*. Kluwer Academic Publishers, Boston, 2002.
6. N.G. Hingorani. Introducing Custom Power. *IEEE Spectrum*, 32(6), 41–48, 1995.
7. J. Dixon, M. Ortuzar, R. Carmi, P. Barriuso, P. Flores, L. Moran. Static Var Compensator and Active Power Filter with Power Injection Capability, Using 27-Level Inverters and Photovoltaic Cells. In *IEEE International Symposium on Industrial Electronics (ISIE'06)*, Montreal, Canada, July 9–13, 2006, vol. 2, pp. 1106–1111.
8. H. Akagi. Active Harmonic Filters. *Proceedings of the IEEE*, 93(12), 2128–2141, 2005.
9. H. Akagi, H. Fujita, S. Yonetani, Y. Kondo. A 6.6-kV Transformerless STATCOM Based on a Five-Level Diode-Clamped PWM Converter: System Design and Experimentation of a 200-V, 10-kVA Laboratory Model. In *IEEE Industry Applications Annual Meeting (IAS'95)*, Orlando, FL, October 8–12, 1995, pp. 557–564.

10. H. Akagi, E.H. Watanabe, M. Aredes. *Instantaneous Power Theory and Applications to Power Conditioning.* IEEE Press, Piscataway, NJ, 2007.
11. H. Akagi, S. Inoue, T. Yoshii. Control and Performance of a Transformerless Cascade PWM STATCOM with Star Configuration. *IEEE Transactions on Industry Applications,* 43(4), 1041–1049, 2007.

chapter 7

Case Study 2: Medium-Voltage Motor Drive Built with DCMC

7.1 Introduction

The use of multilevel converters in electric drives is a topic of great interest, especially when dealing with medium-voltage machines. Besides the main advantage of voltage extension when using multilevel converters, the synthesis of a stepped voltage waveform impacts positively in some key aspects of the conversion system, as, for example, the insulation failures due to common mode voltages and wave reflections, the lifetime reduction of motor bearings due to parasitic capacitive currents, and machine derating due to the harmonic currents on the windings [1]. Due to the lower harmonic content of the stepped voltage waveforms, the multilevel converters are also promising from the grid point of view because they provide power quality improvements that may help to accomplish electrical grid regulations. This applies when power is either drained from the grid (motor drives) or injected to the grid (generators). Medium-voltage drives are commonly used in the oil industry, water treatment, and high-power blowers, among other heavy industrial activities, while medium-voltage generation is gaining great impulse in the renewable power industry, especially for high-power wind turbines. This is due to the necessity to increase the turbine size in order to reach a lower price per unit of generated power [2,3].

Industrial medium-power variable speed drives (VSDs) are commonly based on voltage source inverters in which an AC-to-DC rectification of the sinusoidal input voltage is needed to feed the DC bus. In order to comply with current THD requirements, the DC bus capacitor is usually fed by means of phase-shifting transformers in conjunction with diode rectifiers. Although these circuits allow us to diminish the current harmonics to acceptable values, they involve the use of heavy line frequency transformers that represent a big percentage of the total weight of the converter. On the other hand, feeding the DC bus through an active front-end or back-to-back (B2B) configuration, instead of diode rectifiers, is a very interesting topology that allows the drawing of a nearly sinusoidal current from the grid with a controllable power factor and reduced input filters. Additionally, it provides a way to return mechanical energy to the grid when regenerative braking is required, and also, the boosting capability of the line side converter allows the increase of the DC bus voltage to above the peak line voltage of the power supply.

In this sense, the diode-clamped multilevel converter (DCMC) has a common DC bus, and thus it is inherently suitable for B2B connection. Moreover, the B2B connection of two DCMCs has a major benefit: it establishes a way to potentially extend the DC bus balancing boundary beyond the balancing limits described in Section 3.4. Unfortunately, although the B2B connection of two DCMCs provides a way to compensate the voltage deviations on the DC bus capacitors, it does not ensure the voltage balancing action by itself, so it still needs a closed-loop balancing control that adequately exploits the converter's redundant switching states. On the other hand, apart from the voltage balance issue, the external variables of the B2B converter have to be controlled. Depending on the application, the main control targets may be the AC output current, voltage, power, or even internal variables of the load. This defines a complex control problem with multiple variables (internal and external) that additionally has to include the inherent switching restrictions that were already analyzed in Chapter 3.

In the present chapter, a complete control scheme for a back-to-back DCMC converter in a motor drive application is described. An advanced control method called finite-states model predictive control (FS-MPC) is explained and applied in order to provide a unified control for the voltage balancing of the DC bus and the target variables on the grid and the load side. The complete control problem of two DCMCs operating as the power interface of a medium-voltage induction motor (IM) drive is described. The particular control objectives of the grid converter and the load converter, as well as their solutions, are analyzed and developed. Although the chapter is concentrated in the analysis of a multilevel induction motor drive, the concepts and control philosophy are easily extended to other types of machines and applications, for example, synchronous machines for wind power applications [4].

7.2   Back-to-Back DCMC Converter

In Chapter 3, the voltage drift of the multitapped DC bus of the diode-clamped converter has been analyzed with the basis of a simplified average current flow model and considering the operation in rectifier and inverter mode. It was seen that when the DC bus voltage is fixed and the active power flows from the AC source to the DC bus (rectifier mode), the inner capacitors are charged and the outer capacitors are discharged (Figure 7.1a). The opposite occurs when the power is transferred from the DC bus to the AC source (inverter mode) (Figure 7.1b); that is, the inner

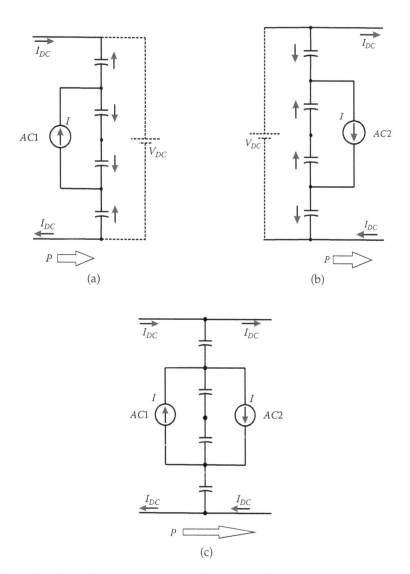

Figure 7.1 Equivalent average current flow model of the DC bus: (a) power flowing from AC1 to the DC bus, (b) power flowing from the DC bus to AC2, and (c) power flowing from AC1 to AC2 through the DC link.

capacitors are discharged and the outer capacitors are charged. If the transferred power and the DC bus voltage V_{DC} are the same in both cases, the discharge tendencies are opposite on both converters, and moreover, the rate of charge/discharge becomes identical. This suggests that if two DCMCs are connected to a common DC bus (Figure 7.1c) and one of them

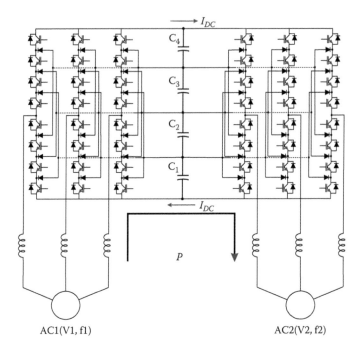

Figure 7.2 Five-level DCMC converter in back-to-back configuration.

injects a power *P* from one AC system (say AC1) to the DC bus while the other extracts the same power from the DC bus and injects it into AC2, a potential counteracting effect can be obtained.

Figure 7.2 shows a frequency converter comprising two five-level DCMCs in back-to-back configuration linking two three-phase systems, AC1 and AC2. This general representation allows us to describe a connection to a three-phase power system with an inductive coupling or may also represent the stator of a machine with its leakage inductance and back electromotive force (EMF). Considering that a given active power *P* flows from AC1 to AC2, the line converter (AC1) works in rectifier mode while the load converter (AC2) works in inverter mode. Assuming that the DC bus voltage is constant, the interaction between both converters of Figure 7.2 becomes evident when associated with Figure 7.1; that is, under certain conditions the power transfer between AC1 and AC2 could be completely achieved through the main bus current I_{DC}, while keeping the balance on the DC bus capacitors. However, this is only true if an adequate managing of the switching states is carried out simultaneously on both converters, since the back-to-back connection of two DCMCs does not self-ensure voltage balance, as was described by Marchesoni and Tenca [5].

In addition to the previous discussion, which strictly regards the internal operation of the back-to-back DCMC, a set of desired features of the B2B system can be determined from the point of view of the application user and are summarized as:

- Operation with arbitrary modulation index and power factors on both converters without mutual dependencies
- Voltage or current control capability at the load side
- Balancing of the DC bus voltages for all operating conditions
- Reactive power control at the grid side
- Low harmonic distortion in the AC currents

The classical control schemes are commonly based on voltage modulators and linear controllers. These structures comprise a linear control loop that calculates the continuous voltage reference that has to be applied to the load in order to achieve the control target. In each period of the carrier signal the continuous voltage reference is converted to logic states of the power switches through pulse width modulation (PWM) or SVM modulation. In this approach, the converter is considered merely as a voltage amplifier that supplies a switched voltage waveform whose average value is an amplified version of the continuous reference signal. However, when inserted in a linear control loop, the switching frequency of the power converter should be sufficiently high in order to avoid interaction with the slower dynamics of the controller. Also, the linear controller is designed for a given point of operation around which a system linearization is accomplished. However, the performance of the controller can be significantly degraded as the operating point moves away from the linearization point, and also, as the system nonlinearities are not considered inside the controllers, their negative effects on the control loops need to be taken into account externally.

In general terms, it can be stated that regardless of the adopted control strategy, the control of internal and external variables of a switching converter resides on the proper selection of its switching states.

7.3 Unified Predictive Controller of the Back-to-Back DCMC in an IM Drive Application

Although the regulation of DC bus voltage balance is a fundamental control goal, the complete control challenge of the back-to-back DCMC also includes the regulation of the external variables on the AC sides of both converters. Such variables depend on the application and may be the AC currents, the active and reactive power, or the machine variables, such as speed, torque, or flux. In this sense, the concept of state optimization via a cost function can be also extended to control other variables that are external to the system. Such a control strategy has been called model predictive control (MPC), and a particular variant called finite-states model

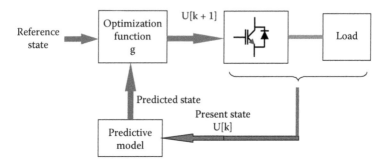

Figure 7.3 Predictive controller block diagram.

predictive control (FS-MPC) has been the subject of intensive research effort related to power converters' control [6–9]. A general block diagram of an FS-MPC predictive controller is shown in Figure 7.3.

At instant k the dynamic state of the system is sampled jointly with the present switching state of the converter. The set of potential switching combinations includes all candidates that can be applied at instant $k + 1$, and each candidate is evaluated by the predictive model. One step-ahead computation is performed such that the output of the predictive model can be compared with the reference state by means of the cost function g. A minimization criterion is chosen for g such that g is minimized when the predicted states approach the reference values. Variables of different types can be evaluated together, providing flexibility to perform multi-variable control schemes. This technique has gained significant popularity for power converters' control, and several publications include current controllers, power controllers, and also the control of rotating machines [8,10–13].

7.3.1 Control of the Back-to-Back DCMC Converter

It has been seen that the capacitor voltage balancing is a fundamental premise to ensure safe operation of the converter, and thus it is an important state that has to be controlled. By using suitable dynamic models, this scheme can be extended to control a set of external variables in a unified control approach by embedding cost functions associated with external variables. This approach allows us to jointly control the internal and external variables altogether through a direct selection of switching states and eliminating the need of modulators and converter-averaged models. As already explained, the control strategy consists in the pre-calculation of target variables from instant k to instant $k + 1$, and the subsequent comparison with their reference values through the cost function (7.1):

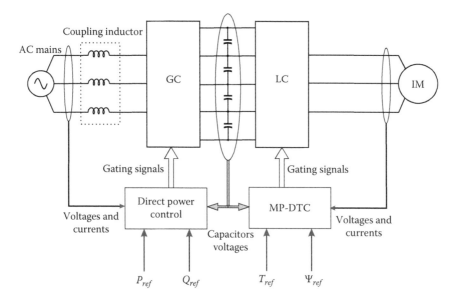

Figure 7.4 Control scheme of a motor drive based on a back-to-back five-level DCMC.

$$g = g_{ext} + g_{int} \tag{7.1}$$

where g_{ext} and g_{int} are the corresponding cost functions of the external and internal variables. The external variables could be, for example, line currents, active and reactive power, and even variables inside the load, while the internal variables are the voltages across the DC bus capacitors. For both DCMC converters of Figure 7.2, the function g_{int} is evaluated for each corresponding set of allowed switching states, through the calculation of capacitor voltages and the evaluation of their mean voltage error according to the algorithm explained in Chapter 3.

$$g_V = \frac{1}{n-1} \sum_{i=1}^{n-1} \frac{\left| V_{Cref} - V_{Ci}[k+1] \right|}{V_{Cref}} = \frac{1}{4} \sum_{i=1}^{4} \frac{\left| V_{Cref} - V_{Ci}[k+1] \right|}{V_{Cref}} \tag{7.2}$$

Figure 7.4 shows a block diagram of the five-level back-to-back DCMC in an induction motor drive application. The converter is fed from a medium-voltage grid without a coupling transformer. The internal and external variables of each converter are controlled jointly through the FS-MPC technique. Specifically, the control goals for the grid converter (GC) are the regulation of the active power for the stabilization of the DC bus voltage and also the reactive and harmonic currents consumed from the grid. On the other hand, the load converter (LC) regulates the flux and the electromagnetic torque through a predictive direct torque control scheme.

An adequate dynamic modeling and the characterization of operational constraints allow us to perform both control tasks jointly with the voltage balancing in an integrated approach. The optimization functions for the grid converter and the load converter can be defined as

$$g_{LC} = g_{LCext} + g_{LCint} = (K_T g_T + K_\psi g_\psi) + (K_{VLC} g_V) \quad \text{(a)}$$

$$\text{(7.3)}$$

$$g_{GC} = g_{GCext} + g_{GCint} = (K_P g_P + K_Q g_Q) + (K_{VGC} g_V) \quad \text{(b)}$$

where g_T and g_ψ are the LC cost functions associated with the electromagnetic (EM) torque and flux regulation, while g_P and g_Q are the GC cost functions associated with the active and reactive power taken from the grid. The coefficients K_i are the weighting factors that allow emphasizing the importance of each term in the optimization process.

7.3.2   Load Converter: Predictive Torque Control

Regarding motor drive controllers, one of the advanced control algorithms for induction motors is the well-known direct torque control. This technique dispenses the use of modulators and external control loops by directly selecting the converter's switching states to steer machine parameters within a preset tolerance band. It achieves this by using a dynamical model of the machine and a decision table that relates the motor flux and EM torque variations, depending on converter switching states. In order to introduce the variant of model predictive direct torque control (MP-DTC) for multilevel converters, a brief review of the classic DTC for two-level inverters is given in the following paragraphs.

The stator winding equation of the induction motor in stationary coordinates is

$$\mathbf{V}_s = r_s \mathbf{i}_s + \frac{d\psi_s}{dt} \quad \text{(7.4)}$$

where $\mathbf{V}_s$, $\mathbf{i}_s$, and ψ_s are the input voltage on motor windings, the stator current, and the stator flux, respectively. By following a simplified analysis in which the voltage drop on r_s is neglected, Equation (7.4) can be approximated through (7.5)

$$\Delta\psi_s \simeq V_s \Delta t \qquad (7.5)$$

That is, if a voltage vector V_s is applied to the stator terminals during a time interval Δt, the flux vector ψ_s will be modified in the same direction as V_s, and also, the modulus of ψ_s will increase proportionally to Δt. Moreover, according to Takahashi and Noguchi [14], under constant flux and rotor speed, the EM torque is proportional to the slip frequency:

$$T_{em} \propto \omega_s \qquad (7.6)$$

This means that the torque can be controlled by accelerating or decelerating the rotating magnetic field. Then, the flux and torque can be controlled by means of flux variations in the radial and tangent directions of the applied voltage vector. Figure 7.5 shows the possible vectors that a two-level inverter can synthesize and six sectors that are defined between converter vectors. Given a flux vector, the application of V_1 along Δt results

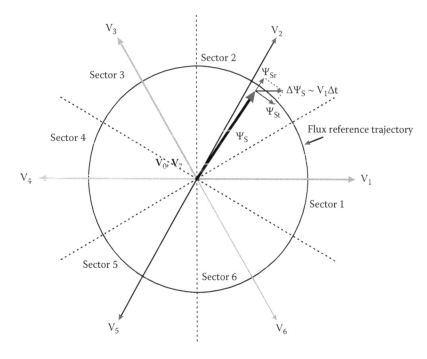

Figure 7.5 Tangent and radial components of flux deviation due to the application of voltage vector V_1 on sector 2.

Table 7.1 DTC Vector Selection Table for a Two-Level VSC

Desired Variation		Target Vector Depending on the Sector					
		1	2	3	4	5	6
$\psi_s\uparrow$	$T_{em}\uparrow$	V_2	V_3	V_4	V_5	V_6	V_1
	$T_{em}=$	V_7	V_0	V_7	V_0	V_7	V_0
	$T_{em}\downarrow$	V_6	V_1	V_2	V_3	V_4	V_5
$\psi_s\downarrow$	$T_{em}\uparrow$	V_3	V_4	V_5	V_6	V_1	V_2
	$T_{em}=$	V_0	V_7	V_0	V_7	V_0	V_7
	$T_{em}\downarrow$	V_5	$V6$	V_1	V_2	V_3	V_4

in a small flux variation with radial and tangent components ψ_{sr}, ψ_{st}, as shown in Figure 7.5. According to the available voltage vector synthesizable by the converter, the current sector in which the tip of the flux vector is located, and the required variation of ψ_s, it is possible to select an adequate vector. Given a flux vector ψ_s, the application of any converter vector during a short period produces flux deviations with different radial and tangent components. Then, different converter vectors can be applied in order to track torque reference and flux reference signals. Table 7.1 shows a basic DTC lookup table that determines a proper converter vector for each combination of the desired variations of T_{em} and ψ_s, depending on the sector in which the flux is located.

Table 7.1 evidences that a direct relationship can be established between the flux and torque variations and the possible vectors that can be synthesized by the two-level converter. Unfortunately, a corresponding selection table for multilevel converters is difficult to be synthesized due to the high number of switching states, especially when the number of levels is high. Then, this mechanism cannot be directly extended to multilevel converters, and a more sophisticated algorithm has to be developed for the selection of converter states [15]. To overcome this difficulty Table 7.1 can be replaced by a suitable cost function that evaluates the converter's most adequate switching state for a joint optimization of flux and torque. This can be achieved through (7.7)

$$g_{LCext} = \left(K_T g_T + K_\psi g_\psi\right) \tag{7.7}$$

where g_T and g_ψ are the cost functions associated with the flux and torque, respectively, and K_T and K_ψ their corresponding weighting factors. Both cost functions are designed to evaluate the difference between the respective command signals with their corresponding predicted values, which are obtained using a dynamic representation of the motor. A prediction model of the IM can be derived from its dynamic equations in fixed stator coordinates:

$$
\left|
\begin{aligned}
&\mathbf{V}_s = r_s \mathbf{i}_s + \frac{d\mathbf{\psi}_s}{dt} &\text{(a)}\\[4pt]
&0 = r_r \mathbf{i}_r + \frac{d\mathbf{\psi}_r}{dt} - jp\omega_m \mathbf{\psi}_r &\text{(b)}\\[4pt]
&\mathbf{\psi}_s = L_s \mathbf{i}_s + L_m \mathbf{i}_r &\text{(c)}\\[4pt]
&\mathbf{\psi}_r = L_m \mathbf{i}_s + L_r \mathbf{i}_r &\text{(d)}\\[4pt]
&T_{em} = \frac{3}{2} p\, \mathrm{Im}\!\left(\mathbf{\psi}_s \mathbf{i}_s^{*}\right) &\text{(e)}
\end{aligned}
\right.
\qquad (7.8)
$$

where L_m, L_r, L_s, r_r, and r_s are the machine electrical parameters, p is the number of pole pairs, and ω_r is the rotor frequency ($\omega_r = p\omega_m$, with ω_m being the rotor speed). As specified by (7.8e), the calculation of electromagnetic torque at a given instant requires the calculation of both the stator flux and the stator current. Considering this, a more consistent representation can be derived from (7.8), which leads to (7.9) [16]:

$$
\left|
\begin{aligned}
&\frac{d\mathbf{\psi}_s}{dt} = \mathbf{V}_s - r_s \mathbf{i}_s &\text{(a)}\\[4pt]
&\tau_{sr}' \frac{d\mathbf{i}_s}{dt} + \mathbf{i}_s = j\omega_r \tau_{sr}' \mathbf{i}_s + \frac{1}{r_{sr}}\left(\frac{1}{\tau_r} - j\omega_r\right)\mathbf{\psi}_s + \frac{1}{r_{sr}}\mathbf{V}_s &\text{(b)}
\end{aligned}
\right.
\qquad (7.9)
$$

where

$$
\tau_{sr}' = \frac{\sigma L_s}{r_{sr}}, \quad r_{sr} = r_s + \frac{L_s}{L_r} r_r, \quad \tau_r = \frac{L_r}{r_r}, \quad \sigma = 1 - k_s k_r, \quad k_s = \frac{L_m}{L_s}, \quad k_r = \frac{L_m}{L_r}
$$

First, the forward calculation of the stator flux can be achieved by discretizing the flux derivative in (7.9a). However, the pure integration of (7.9a) and the inability to sample the actual value of the flux leads to cumulative errors, whereby its direct discretization is avoided. Instead, the integrating operator is replaced by a low-pass filter with a suitable cutoff frequency ω_{co}, as shown in Figure 7.6.

The flux estimation can be expressed in the frequency domain through:

$$
\frac{\mathbf{V}_s - r_s \mathbf{i}_s}{s + \omega_{co}} = \mathbf{\psi}_s \quad \Rightarrow \quad \mathbf{V}_s - r_s \mathbf{i}_s = s\mathbf{\psi}_s + \omega_{co}\mathbf{\psi}_s
\qquad (7.10)
$$

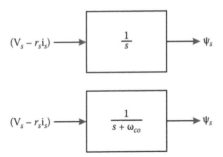

Figure 7.6 Integral operator replaced by a low-pass filter to estimate the stator flux.

Discretizing the derivatives by forward Euler approximation and considering a sampling period T_S, we obtain

$$\mathbf{V}_s[k] - r_s\mathbf{i}_s[k] = \frac{\psi_s[k+1] - \psi_s[k]}{T_S} + \omega_{co}\psi_s[k]$$

$$\Rightarrow \psi_s[k+1] = \left(\mathbf{V}_s[k] - r_s\mathbf{i}_s[k] - \omega_{co}\psi_s[k]\right)T_S + \psi_s[k] \qquad (7.11)$$

$$\Rightarrow \psi_s[k+1] = \left(\mathbf{V}_s[k] - r_s\mathbf{i}_s[k]\right)T_S + \psi_s[k]\left(1 - T_S\omega_{co}\right)$$

Equation (7.11) allows us to calculate the future values of the flux components in terms of its previous prediction, the measured currents, and the applied voltage.

$$\begin{cases} \psi_{s\alpha}^p[k+1] = (V_{s\alpha}[k] - r_s i_{s\alpha}[k])T_S + \left(1 - T_S\omega_{co}\right)\psi_{s\alpha}^p[k] \\ \psi_{s\beta}^p[k+1] = (V_{s\beta}[k] - r_s i_{s\beta}[k])T_S + \left(1 - T_S\omega_{co}\right)\psi_{s\beta}^p[k] \end{cases} \qquad (7.12)$$

where the superscript p denotes predicted variables. Given that the flux cannot be sampled, (7.12) uses the predicted value calculated in the previous sampling interval as an estimation.

In order to make a forward calculation of the EM torque, a prediction of the motor current is needed. The computation of the stator current can be made through the Euler forward discretizing of (7.9b):

$$\tau'_{sr}\frac{\mathbf{i}_s[k+1]-\mathbf{i}_s[k]}{T_S}+\mathbf{i}_s[k]=j\omega_r\tau'_{sr}\mathbf{i}_s[k]+\frac{1}{r_{sr}}\left(\frac{1}{\tau_r}-j\omega_r\right)\mathbf{\psi}_s[k]+\frac{1}{r_{sr}}\mathbf{V}_s[k]$$

$$\Rightarrow\quad \frac{\tau'_{sr}}{T_S}\mathbf{i}_s[k+1]-\frac{\tau'_{sr}}{T_S}\mathbf{i}_s[k]=\mathbf{i}_s[k]\left(j\omega_r\tau'_{sr}-1\right)+\frac{1}{r_{sr}}\left(\frac{1}{\tau_r}-j\omega_r\right)\mathbf{\psi}_s[k]+\frac{1}{r_{sr}}\mathbf{V}_s[k]$$

$$\Rightarrow\quad \mathbf{i}_s[k+1]=\left(\mathbf{i}_s[k]\left(\frac{\tau'_{sr}-T_S}{\tau'_{sr}}+j\omega_r T_S\right)\right)+\frac{T_S}{\tau'_{sr}r_{sr}}\left(\frac{1}{\tau_r}-j\omega_r\right)\mathbf{\psi}_s[k]+\frac{T_S}{\tau'_{sr}r_{sr}}\mathbf{V}_s[k]$$

Decomposing in real and imaginary components it results in

$$
\begin{cases}
i^p_{s\alpha}[k+1]=i_{s\alpha}[k]\dfrac{\tau'_{sr}-T_S}{\tau'_{sr}}-i_{s\beta}[k]\omega_r T_S+\psi^p_{s\alpha}[k]\dfrac{T_S}{\tau'_{sr}r_{sr}\tau_r}\\[1.5em]
\qquad -\psi^p_{s\beta}[k]\dfrac{T_S\omega_r}{\tau'_{sr}r_{sr}}-\dfrac{T_S}{\tau'_{sr}r_{sr}}V_{s\alpha}[k]\\[1.5em]
i^p_{s\beta}[k+1]=i_{s\alpha}[k]\omega_r T_S+i_{s\beta}[k]\dfrac{\tau'_{sr}-T_S}{\tau'_{sr}}+\psi^p_{s\beta}[k]\dfrac{T_S}{\tau'_{sr}r_{sr}\tau_r}\\[1.5em]
\qquad +\psi^p_{s\alpha}[k]\dfrac{T_S\omega_r}{\tau'_{sr}r_{sr}}+\dfrac{T_S}{\tau'_{sr}r_{sr}}V_{s\beta}[k]
\end{cases}
\qquad (7.13)
$$

Equations (7.12) and (7.13) allow us to precalculate both the square of the flux modulus and the EM torque at instant $k + 1$ using the sampled and estimated variables at instant k for all possible combinations of $\mathbf{V}_s[k]$:

$$
\begin{cases}
\left(\psi^p_s[k+1]\right)^2=\left(\psi^p_{s\alpha}[k+1]\right)^2+\left(\psi^p_{s\beta}[k+1]\right)^2 & \text{(a)}\\[1.5em]
T^p_{em}[k+1]=\dfrac{3}{2}p\left(\psi^p_{s\alpha}[k+1]i^p_{s\beta}[k+1]-\psi^p_{s\beta}[k+1]i^p_{s\alpha}[k+1]\right) & \text{(b)}
\end{cases}
\qquad (7.14)
$$

The corresponding cost functions can be designed similarly as (7.2) for the comparison of torque and flux with the reference values

$$g_T=\frac{\left|T_{ref}-T^p_{em}\right|}{T_{ref}}\quad\text{and}\quad g_\psi=\frac{\left|\psi^2_{ref}-(\psi^p_s)^2\right|}{\psi^2_{ref}}\qquad (7.15)$$

The reference value ψ_{ref} is fixed according to the rated voltage and line frequency of the machine, while the torque reference is set by the external speed controller.

In each sampling period, the evaluation of g_T and g_ψ is made jointly with g_V for the allowed switching combinations, and the one that minimizes (7.3a) is selected as the next state of the LC.

7.3.3   Line Converter: Predictive Power Control

The control goal of the GC is the regulation of the DC voltage and the reactive power demanded from the grid. The associated cost function can be expressed as

$$g_{GCext} = K_P g_P + K_Q g_Q \tag{7.16}$$

The function g_{GCext} evaluates the converter switching combinations in order to take the necessary active power from the grid to regulate the DC bus voltage, and also to minimize the reactive power consumption.

7.3.3.1   Current and Power Calculations

The line converter is connected to a three-phase grid through the coupling inductor as depicted in Figure 7.7. Given that the system is three-wired without a neutral conductor, the converter currents are determined from the line voltages e_{ab}, e_{bc}, V_{ab}, and V_{bc}, where the last two assume integer multiples of capacitor voltage V_C from $-(n-1)V_C$ to $(n-1)V_C$ (in this case $n = 5$). The circuit equations are

$$\begin{cases} V_{ab} - L\dfrac{di_a}{dt} - Ri_a - e_{ab} + L\dfrac{di_b}{dt} + Ri_b = 0 \\[2mm] V_{bc} - L\dfrac{di_b}{dt} - Ri_b - e_{bc} + L\dfrac{di_c}{dt} + Ri_c = 0 \\[2mm] i_a + i_b + i_c = 0 \end{cases} \tag{7.17}$$

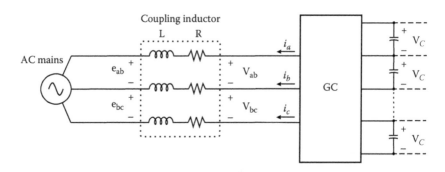

Figure 7.7 DCMC connection to the grid through a coupling inductor.

Using the forward Euler approximation for the derivatives, (7.17) can be written as

$$
\begin{cases}
V_{ab}[k] - L\left(\dfrac{i_a[k+1]-i_a[k]}{T_S}\right) - Ri_a[k] - e_{ab}[k] + L\left(\dfrac{i_b[k+1]-i_b[k]}{T_S}\right) + Ri_b[k] = 0 \\[2mm]
V_{bc}[k] - L\left(\dfrac{i_b[k+1]-i_b[k]}{T_S}\right) - Ri_b[k] - e_{bc}[k] + L\left(\dfrac{i_c[k+1]-i_c[k]}{T_S}\right) + Ri_c[k] = 0 \\[2mm]
i_c[k] = -(i_a[k]+i_b[k])
\end{cases}
$$

Substituting i_c,

$$
\begin{cases}
V_{ab}[k] - e_{ab}[k] - L\dfrac{i_a[k+1]-i_a[k]}{T_S} - Ri_a[k] + L\dfrac{i_b[k+1]-i_b[k]}{T_S} + Ri_b[k] = 0 \\[2mm]
V_{bc}[k] - e_{bc}[k] - 2L\dfrac{i_b[k+1]-i_b[k]}{T_S} - 2Ri_b[k] - L\dfrac{i_a[k+1]-i_a[k]}{T_S} - Ri_a[k] = 0
\end{cases}
$$

solving for $i_a[k + 1]$ leads to

$$
i_a[k+1] = \frac{T_S}{3L}\left(2V_{ab}[k]+V_{bc}[k]-2e_{ab}[k]-e_{bc}[k]\right)+\left(1-\frac{RT_S}{L}\right)i_a[k] \quad (7.18)
$$

Solving for i_b and i_c, the three currents can be simultaneously expressed:

$$
\begin{bmatrix} i_a^p \\ i_b^p \\ i_c^p \end{bmatrix}_{K+1} = \frac{T_S}{3L}\left(\begin{bmatrix} 2 & 1 \\ -1 & 1 \\ -1 & -2 \end{bmatrix}\left(\begin{bmatrix} V_{ab} \\ V_{bc} \end{bmatrix}_K - \begin{bmatrix} e_{ab} \\ e_{bc} \end{bmatrix}_K\right)\right)+\left(1-\frac{RT_S}{L}\right)\begin{bmatrix} i_a \\ i_b \\ i_c \end{bmatrix}_K \quad (7.19)
$$

At instant k the system voltage $\mathbf{e}[k]$ and current $\mathbf{i}[k]$ are sampled. Then, for each set of line voltages at the output of the converter, the instantaneous active and reactive power can be precalculated to the instant $(k + 1)$, considering that the system voltage has slow variations within a sampling period $(\mathbf{e}[k] \approx \mathbf{e}[k + 1])$:

$$
P^p[k+1] = e_\alpha[k]i_\alpha^p[k+1] + e_\beta[k]i_\beta^p[k+1]
$$
$$
Q^p[k+1] = e_\beta[k]i_\alpha^p[k+1] - e_\alpha[k]i_\beta^p[k+1]
$$

$$(7.20)$$

The cost functions for P and Q are defined in (7.21) and account for both power errors.

$$g_P = \left| \left(\frac{P_{ref} - P^p}{P_{ref}} \right) \right| \quad \text{and} \quad g_Q = \left| \left(\frac{Q_{ref} - Q^p}{P_{ref}} \right) \right| \tag{7.21}$$

The reference Q_{ref} is forced to zero in order to achieve the unity power factor of the grid current. On the other hand, given that the DC bus voltage variations are linked to the power flow between the GC and the LC, the active power reference signal P_{ref} can be determined from an estimation of the power consumed by the load converter and a term related to the voltage error of the DC bus voltage [17].

7.3.3.2 *Dynamic Active Power Reference Design*

The necessary power that has to be taken from the grid for the stabilization of the DC bus voltage is calculated as the sum of a feed-forward term that accounts for the power delivered to the load and a feedback term that evaluates the voltage error of the DC bus with respect to its reference value. If a power mismatch occurs between the grid converter and the load converter (Figure 7.8), for example, if the LC draws more power than the one that is taken from the grid by the GC, the power difference will be supplied by the DC bus. In this condition the DC bus voltage will decrease, and the GC will have to inject an additional amount of energy to return it to its reference value. The energy balance equation is

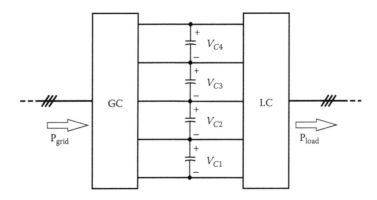

Figure 7.8 Power balance between the grid converter and the load converter.

$$P_{grid} - P_{load} = P_C = \frac{dE_C}{dt} = \frac{d}{dt}\left(\sum_{i=1}^{4} E_{Ci}\right) = \frac{d}{dt}\left(\sum_{i=1}^{4} \frac{1}{2}CV_{Ci}^2\right)$$

$$= \frac{1}{2}C\left(\sum_{i=1}^{4} 2V_{Ci}\frac{dV_{Ci}}{dt}\right) \approx C\frac{V_{DC}}{4}\frac{\Delta V_{DC}}{T_S} \tag{7.22}$$

The additional power that should be injected to the capacitor's bank is then

$$P_C = C\frac{V_{DC}}{4}\frac{\Delta V_{DC}}{T_S} = \frac{CV_{DC}}{4T_S}\left(V_{DCref} - V_{DCmeas}\right) \tag{7.23}$$

Equation (7.23) is a corrective term that directly monitors the DC bus voltage. The active power consumed by the LC can be estimated and added to the active power reference signal in order to improve the regulation dynamics. This estimation can be determined using the EM torque (7.14b) and rotor speed, such that

$$P_{load} = T_{em}\omega_m \tag{7.24}$$

The reference power P_{ref} for the evaluation of g_p can be calculated as

$$P_{ref} = P_{grid} = T_{em}\omega_m + \frac{CV_{DC}}{4T_S}\left(V_{DCref} - V_{DCmeas}\right) \tag{7.25}$$

7.3.4 Switching Transition Constraint

The search of the switching combination that minimizes (7.3a) and (7.3b) involves a hard computational effort. Although the control algorithms can be executed in parallel, a high number of calculations have to be processed, and this gets worse as the number of voltage levels increases. In the case of the five-level converter, the total amount of available states is 5^3 = 125. In addition, all the calculations have to be executed in a short time due to the high-frequency switching of the power devices. However, as seen in Section 3.1, the switching transition cannot be made from a given state to any arbitrary state, because multiple step transitions are not permitted on leg voltages. This is so because such transitions do not ensure a safe clamping of blocking voltages in the power switches, whereby only some states are allowed. Then, a given switching state can be only

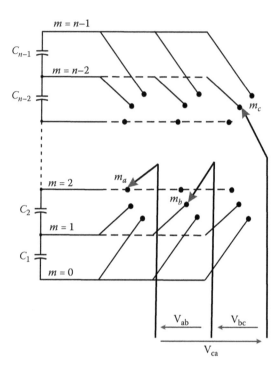

Figure 7.9 Functional model of the three-phase *n*-level DCMC.

succeeded by another state that does not produce multiple step transitions in any leg of the converter, shortening the set of possible switching states for the calculations. To introduce the way this is taken into account, the functional model of the *n*-level DCMC is presented in Figure 7.9. During the sampling period *k*, the three legs are connected to the corresponding taps m_a, m_b, m_c (m_a, m_b, $m_c = 0, ..., n - 1$), thus defining the line voltages V_{ab}, V_{bc}, and consequently V_{ca}. By virtue of the single-step switching transition of Section 3.1, we can specify a set of admissible positions (*Sw*) for each leg at time instant *k* + 1 in which Δ_x ($x = a, b, c$) represents the switch displacement on each leg of the converter.

$$Sw[k+1]: \begin{cases} m_a[k+1] = m_a[k] + \Delta_a & 0 \le m_a \le n-1 \text{ with } \Delta_a = 0,1,-1 \\ m_b[k+1] = m_b[k] + \Delta_b & 0 \le m_b \le n-1 \text{ with } \Delta_b = 0,1,-1 \quad (7.26) \\ m_c[k+1] = m_c[k] + \Delta_c & 0 \le m_c \le n-1 \text{ with } \Delta_c = 0,1,-1 \end{cases}$$

Considering all possible combinations of voltage variations of Δ_a, Δ_b, and Δ_c, it is clear that the subset *Sw* has a maximum number of elements

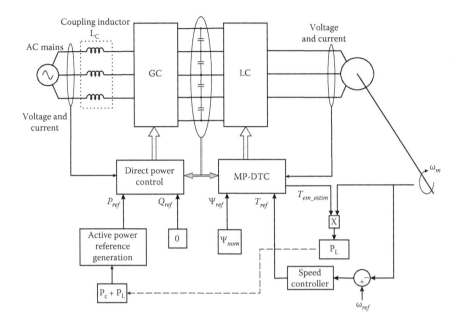

Figure 7.10 Back-to-back DCMC converter driving an IM.

equal to $3^3 = 27$, independently of the number of levels of the converter. In addition, this number may be further reduced as the modulation index approaches unity because of the limits imposed also on m_a, m_b, and m_c $0 \leq m_x \leq (n-1)$. As the calculation of the predicted variables is made for the allowed switching states, this constraint relaxes the computational requirements. In particular, for the five-level DCMC, the total number of switching states is decreased from 125 to 27, which is a significant reduction in computational effort.

7.4 Performance Evaluation

The back-to-back DCMC converter was configured to drive a medium-voltage (MV) asynchronous motor whose mechanical load consists of a blower with a quadratic torque-speed characteristic. The complete system is shown in Figure 7.10. The line converter is connected to the grid through the coupling reactor L_C, while the load converter drives the IM directly. The machine's rated voltage is 13 kV, while the DC bus voltage is regulated to 20 kV. The GC controller gets its active power reference P_{ref} from the computation of the estimated mechanical power delivered to the load P_L (7.24) and the DC bus regulation P_C (7.23). In every sampling period, the necessary active power that has to be taken from the grid is determined. The reactive power reference signal Q_{ref} is forced to zero in order to ensure that a high power factor is presented to the grid. On the other hand, the reference signals for the LC controller are the rated flux of the IM, which can be determined from machine specification sheet and also the EM torque reference. In this case study, an external speed controller commands the EM torque reference to regulate the rotational speed of the load. It is worth mentioning that there is a small interaction between both controllers, since only the estimated load power (P_{load}) is interchanged between the GC and LC controllers. This means that there is no interdependence on the selection of the switching states on both converters such that both control routines can be executed independently on different processors, or moreover, they can be executed in parallel.

In order to examine the behavior of the controlled variables, four tests are proposed:

1. Sudden change of mechanical load on the motor shaft
2. Voltage sag on the grid voltage
3. Energy recovery through regenerative braking
4. Evaluation of the DC bus voltage balance due to the associated cost functions g_{LCint} and g_{GCint} and also the compensation effect due to the B2B connection

7.4.1 Mechanical Load Variation

Figure 7.11 shows the electrical variables on the line converter when the load is disconnected at $t = 2$ s. Before this moment, the blower exerts a load torque of approximately 3600 Nm, and it is suddenly decreased to zero. One line voltage on the grid side and its counterpart on the converter output are shown superimposed in Figure 7.11a. The trace of converter line voltage is observed with single-voltage jumps providing a soft voltage waveform even at the instant of load torque perturbation. On the other hand, the amplitude of all voltage levels is kept constant along time without any drift, whereby the effectiveness of the DC bus voltage balance is verified. Previous to the perturbation, a small steady-state error on the DC bus voltage can be seen (Figure 7.11b), while this value increases to approximately 0.25% during the transient and is finally restored in approximately 40 ms. The current drained from the grid by the GC is almost sinusoidal, has low harmonic content during power consumption, and is decreased to zero when mechanical load is disaffected (Figure 7.11c). Figure 7.11d shows one line voltage of the load converter and one current supplied to the IM. A higher harmonic content is observed in this case than with the grid converter. However, single jumps are observed in almost all the traces, while the current has a sinusoidal profile whose amplitude and phase shift are modified at the instant of the mechanical load decoupling. The current supplied to the motor after the

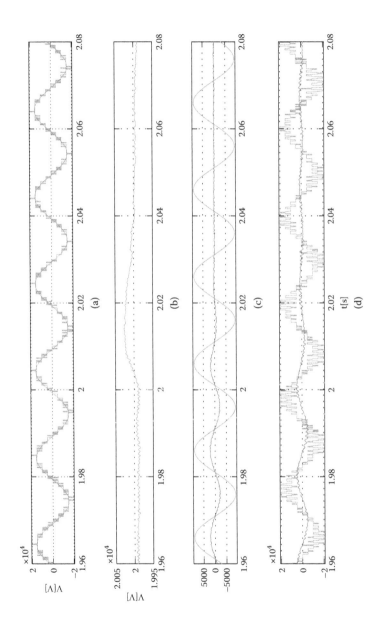

Figure 7.11 (a) Line voltages at both sides of the coupling inductor L_C. (b) DC bus voltage. (c) Phase voltage and its corresponding current at the grid side. (d) Line voltage and current on motor terminals.

load disconnection is, of course, the magnetization current that regulates the flux to the reference value. Figure 7.12a shows the trajectory described by the flux vector, and Figure 7.12b depicts the reference and the measured rotor speed, ω_m. First, the speed value is close to the reference value. At this point, the power supplied to the rotor is 540 kW until $t = 2$ s, when the mechanical torque is nullified (Figure 7.12c).

7.4.2   Voltage Sag

The second test consists of a transient reduction of power system voltage. Figure 7.13 shows the electrical variables when a voltage sag with an amplitude of 25% occurs in $t = 2$ s. Figure 7.13a shows the line voltage at both sides of the coupling inductor. The amplitude reduction on the system voltage can be observed, and also the fundamental component of the voltage waveform through a decrease of the modulation index. This is so because the DC bus voltage remains fixed at 20 kV. Figure 7.13b shows the DC bus voltage, which does not present any variation. This is due to the fact that the DC bus voltage regulation involves the direct determination of the active power that has to be taken from the grid in order to keep the energy balance between the incoming power, the delivered power to the load, and the energy stored on the DC bus. One phase current and its corresponding phase voltage are shown in Figure 7.13c. It can be seen that the increment of the current, which counteracts the line voltage decrement in order to preserve the power level.

Figure 7.14 shows the capacitor voltages, which present reduced deviations with respect to the reference value of 5 kV during the voltage sag. This is the consequence of a higher availability of redundant states due to the reduction of the modulation index, which increases the number of redundant states of the GC.

7.4.3   Energy Recovery

Figures 7.15 and 7.16 show the result of a sign inversion of the torque reference signal, T_{ref}, due to a reduction of the speed command signal on the external speed controller.

This change is realized at $t = 2$ s. During the time the shaft speed decreases to its new reference value a braking control action is carried out and the energy in excess, which is stored in the rotational mass, is reinjected to the grid at synchronous frequency. In Figure 7.15 one phase voltage and its phase current are shown. The phase shift is rapidly reached passing from power consumption to power generation in just one cycle of the grid voltage. A detail of capacitor voltages is shown in Figure 7.16, where the balance is preserved by the balancing algorithm, even in presence of this severe perturbation.

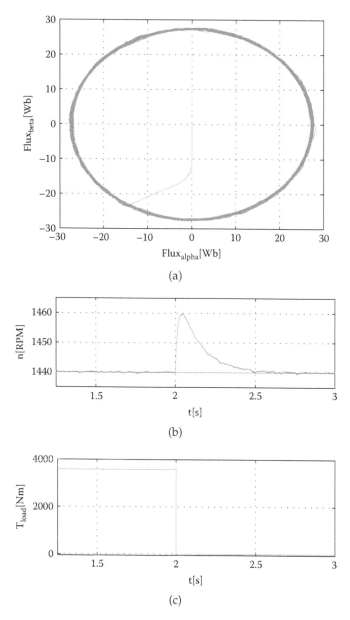

(a)

(b)

(c)

Figure 7.12 (a) Trajectory of magnetic flux vector. (b) Shaft speed at the moment of load torque perturbation. (c) Load torque.

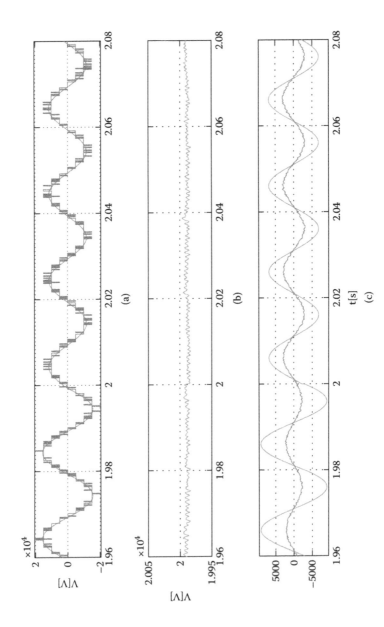

Figure 7.13 (a) Line voltage at both sides of the coupling inductor L_C. (b) DC bus voltage. (c) Phase voltage and corresponding phase current.

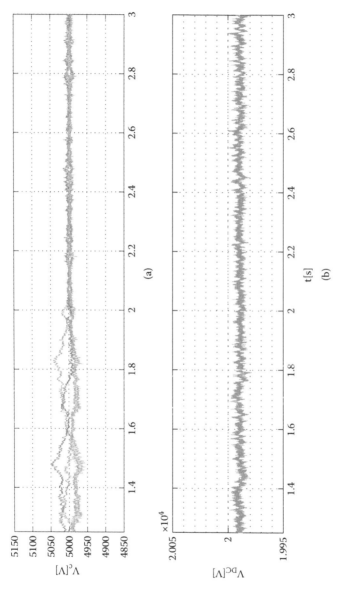

Figure 7.14 (a) Capacitors' voltages. (b) DC bus voltage.

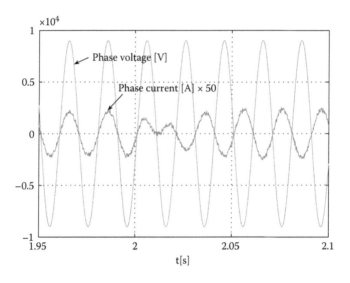

Figure 7.15 Phase voltage and phase current with regenerative braking.

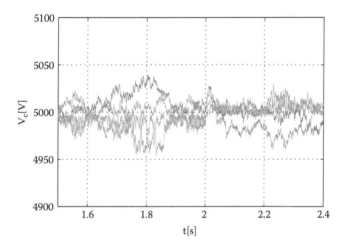

Figure 7.16 Capacitor voltages during regenerative braking in $t = 2$ s.

7.4.4   *Effectiveness of the DC Bus Balancing Algorithm*

In this test, the effectiveness of the cost functions associated with voltage balance for the LC and GC are separately examined. Also, the capacitor voltage balancing effect due to the back-to-back connection is verified. First, the incidence of the cost functions for voltage balance control on both converters is explored. This is achieved by making zero their respective

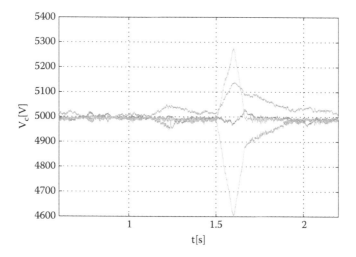

Figure 7.17 Capacitor voltages during $K_{VGC} = 0$.

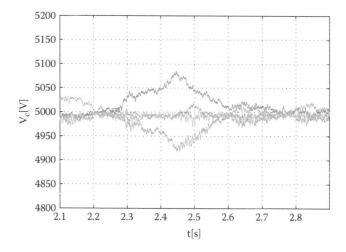

Figure 7.18 Capacitor voltages during $K_{VLC} = 0$.

weighting factors, which clearly opens the voltage balance control loop, according to the minimization criterion explained in Section 7.3.

At first, the system works at the rated conversion level. From $t = 1.5$ s to $t = 1.65$ s, K_{VGC} is modified from its original value 0.1 to 0 in order to disaffect the contribution of the GC to the balancing of the DC bus. Figure 7.17 shows that the capacitor voltages start to diverge from the reference value of 5 kV until K_{VGC} is restored to its original value, at which they begin to return to their original values. Similarly, Figure 7.18 shows the result of the same operation over K_{VLC} from $t = 2.25$ s to $t = 2.45$ s. The result is in

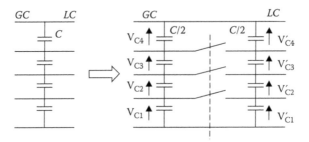

Figure 7.19 Intermediate nodes disconnection.

this case similar to the aforementioned; that is, a divergent evolution is verified for all capacitor voltages that is restored after K_{VLC} returns to its original value. This confirms the fact that the B2B connection does not self-ensure the capacitor voltage balance, and that an optimal selection of the switching combinations on both converters is necessary to keep the DC bus in a balanced condition.

Finally, the compensating effect on voltage balance due to the B2B connection is tested by maintaining the original values of K_{VLC} and K_{VGC} but now opening the interconnection of the internal nodes of the DC bus between the GC and the LC. The test aims to probe that the compensation effect over the DC bus capacitors is possible even when both converters cannot individually maintain the balance. Recalling Figure 7.1, the test forces the pass from Figure 7.1c (B2B connection enabled) to Figure 7.1a,b, in which both converters only share the main DC bus conductors but the intermediate nodes are not linked. This change is made by splitting the DC bus capacitors in two sets of value $C/2$ and associated with each converter. The intermediate nodes are connected through a three-point controllable switch (Figure 7.19) that allows us to control the connection of the intermediate nodes as desired. First, the switch is closed and normal operation is reached. At $t = 2.25$ s the switch is open and reclosed in $t = 2.30$ s. During this time, both converters operate with their respective balancing control activated and the DC bus voltage is regulated to 20 kV by the grid converter. Figure 7.20 shows that immediately after the switch is opened, the capacitor voltages start to diverge in opposite directions for both converters and at the same rate. The opposite charge/discharge on the inner and outer capacitors on the LC and GC can also be observed. The discharge of C_2 and C_3 can be observed on the LC side, which is the typical pattern for power flow direction from the DC side to the AC side ($P_{DC} \rightarrow P_{AC}$), while the opposite occurs with their counterparts on the GC ($P_{AC} \rightarrow P_{DC}$). As was mentioned, the DC bus voltage remains constant due to the active power controller executed by the grid converter, which is independent of both balancing algorithms.

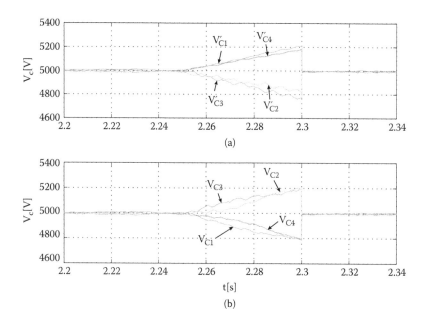

Figure 7.20 Capacitor voltage unbalance due to a temporary disconnection of the intermediate nodes of the DC bus: (a) load converter and (b) grid converter.

It can be concluded from this result that the back-to-back connection of two diode-clamped converters provides a way to extend the balancing boundary of the topology, but it is only effective if it is combined with an intelligent selection of the switching states on both converters.

7.5   Summary

This chapter presents the DCMC in back-to-back configuration. The general benefits of the topology are described from the points of view of both external features and its ability to extend the balancing boundary. A medium-voltage motor drive that uses two five-level DCMCs has been analyzed in which the predictive balancing scheme was extended to control the external variables on both converter sides: the instantaneous power on the grid converter and torque/flux control on the load converter. The inherent switching restrictions are explicitly taken into account in the control algorithm, which ensures safe switching conditions on converter legs and also provides a significant reduction of the computational effort. The integrated control strategy of voltage balancing and external variables has proven to be efficiently carried out through the FS-MPC control algorithm, providing a consistent controller of the target variables and also avoiding the use of external modulators. Simulation results indicate that both the circuit configuration and the control strategy comply with the basic operational features that a back-to-back converter has to exhibit, broadening the usage possibilities to other applications of frequency conversion or power processing, especially at MV levels.

References

1. Bearing Currents in AC Drive Systems, ABB Technical Guide 5. http://www05.abb.com.
2. Z. Chen, J. Guerrero, F. Blaajberg. A Review of the State of the Art of Power Electronics for Wind Turbines. *IEEE Transactions on Power Electronics*, 24(8), 1859–1875, 2009.
3. J.M. Carrasco, L.G. Franquelo, J.T. Bialasiewicz, E. Galvan, R.C.P. Guisado, M.M. Prats, J.I. Leon, N. Moreno-Alfonso. Power-Electronic Systems for the Grid Integration of Renewable Energy Sources: A Survey. *IEEE Transactions on Industrial Electronics*, 53(4), 1002–1016, 2006.
4. S. Alepuz, S. Busquets-Monge, J. Bordonau, P. Cortés, S. Kouro. Control Methods for Low Voltage Ride-Through Compliance in Grid-Connected NPC Converter Based Wind Power Systems Using Predictive Control. In *IEEE Energy Conversion Congress and Exposition (ECCE'09)*, San José, CA, September 20–24, 2009, pp. 363–369.
5. M. Marchesoni, P. Tenca. Diode-Clamped Multilevel Converters: A Practicable Way to Balance DC-Link Voltages. *IEEE Transactions on Industrial Electronics*, 49(4), 752–765, 2002.
6. S. Kouro, P. Cortés, R. Vargas, U. Ammann, J. Rodríguez. Model Predictive Control—A Simple and Powerful Method to Control Power Converters. *IEEE Transactions on Industrial Electronics*, 56(6), 1826–1838, 2009.
7. P. Cortés, M. Kazmierkowski, R. Kennel, D. Quevedo, J. Rodríguez. Predictive Control in Power Electronics and Drives. *IEEE Transactions on Industrial Electronics*, 55(12), 4312–4324, 2008.
8. P. Cortés, J. Rodriguez, P. Antoniewicz, M. Kazmierkowski. Direct Power Control of an AFE Using Predictive Control. *IEEE Transactions on Power Electronics*, 23(5), 2516–2523, 2008.
9. R. Vargas, P. Cortés, U. Ammann, J. Rodríguez, J. Pontt. Predictive Control of a Three-Phase Neutral-Point-Clamped Inverter. *IEEE Transactions on Industrial Electronics*, 54(5), 2697–2705, 2007.
10. J. Rodríguez, J. Pontt, C. Silva, P. Correa, P. Lezana, P. Cortés, U. Ammann. Predictive Current Control of a Voltage Source Inverter. *IEEE Transactions on Industrial Electronics*, 54(1), 495–503, 2007.
11. A. Bouafia, J.P. Gaubert, F. Krim. Predictive Direct Power Control of Three-Phase Pulsewidth Modulation (PWM) Rectifier Using Space-Vector Modulation (SVM). *IEEE Transactions on Power Electronics*, 25(1), 228–236, 2010.
12. P. Correa, M. Pacas, J. Rodriguez. Predictive Torque Control for Inverter-Fed Induction Machines. *IEEE Transactions on Industrial Electronics*, 54(2), 1073–1079, 2007.
13. E. Fuentes, J. Rodriguez, C. Silva, S. Díaz, D. Quevedo. Speed Control of a Permanent Magnet Synchronous Motor Using Predictive Current Control. In *The International Power Electronics and Motion Control Conference (IPEMC'09)*, Wuhan, China, May 17–20, 2009, pp. 390–395.

14. I. Takahashi, T. Noguchi. A New Quick-Response and High-Efficiency Strategy for an Induction Motor. *IEEE Transactions on Industry Applications*, 22(5), 820–827, 1986.

15. S. Kouro, R. Bernal, C. Silva, J. Rodriguez, J. Pontt. High Performance Torque and Flux Control for Multilevel Inverter Fed Induction Motors. In *IEEE Annual Conference of the Industrial Electronics Society (IECON'06)*, Paris, France, November 6–10, 2006, pp. 805–810.

16. J. Holtz. The Representation of AC Machine Dynamics by Complex Signal Flow Graphs. *IEEE Transactions on Industrial Electronics*, 42(3), 263–271, 1995.

17. D. Quevedo, R.P. Aguilera, M.A. Pérez, P. Cortés. Finite Control Set MPC of an Active Rectifier with Dynamic References. In *IEEE International Conference on Industrial Technology (ICIT'10)*, Viña del Mar-Valparaiso, Chile, March 14–17, 2010, pp. 1245–1250.

Index